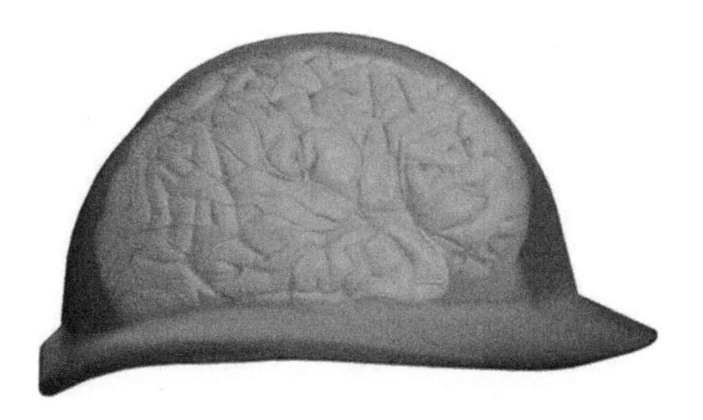

Human Behavior in Hazardous Situations

This page intentionally left blank

Human Behavior in Hazardous Situations

Best Practice Safety Management in the Chemical and Process Industries

Juni Daalmans
Daalmans Organizational Development
www.brainbasedsafety.com
post@daalmans.nl

AMSTERDAM • BOSTON • HEIDELBERG • LONDON
NEW YORK • OXFORD • PARIS • SAN DIEGO
SAN FRANCISCO • SINGAPORE • SYDNEY • TOKYO
Butterworth-Heinemann is an imprint of Elsevier

Butterworth-Heinemann is an imprint of Elsevier
The Boulevard, Langford Lane, Kidlington, Oxford, OX5 1GB, UK
225 Wyman Street, Waltham, MA 02451, USA

First published 2013

Notices

Knowledge and best practice in this field are constantly changing. As new research and experience broaden our understanding, changes in research methods, professional practices, or medical treatment may become necessary.

Practitioners and researchers must always rely on their own experience and knowledge in evaluating and using any information, methods, compounds, or experiments described herein. In using such information or methods they should be mindful of their own safety and the safety of others, including parties for whom they have a professional responsibility.

To the fullest extent of the law, neither the Publisher nor the authors, contributors, or editors, assume any liability for any injury and/or damage to persons or property as a matter of products liability, negligence or otherwise, or from any use or operation of any methods, products, instructions, or ideas contained in the material herein.

British Library Cataloguing-in-Publication Data
A catalogue record for this book is available from the British Library

Library of Congress Cataloging-in-Publication Data
A catalog record for this book is available from the Library of Congress

ISBN: 978-0-12-407209-1

For information on all Butterworth-Heinemann publications
visit our website at store.elsevier.com

This book has been manufactured using Print On Demand technology. Each copy is produced to order and is limited to black ink. The online version of this book will show color figures where appropriate.

DEDICATION

Special thanks to Olaf Derkx, for his initial discussions and support to start the project that led to this book, to Linda Wright for her feedback concerning the depth and the breadth of the theory, to Henk Gerards for his work in constructing a case, and Mark Williams for his contributions to the topic of priming.

This page intentionally left blank

CONTENTS

GENERAL INTRODUCTION TO THIS BOOK

APPROACH AND MAIN QUESTIONS

What should be the starting point of a new safety approach? All theories developed thus far were begun by analyzing previous accidents and casualties. In doing so, safety implicitly is defined as not acting unsafely. Suppose we wish to attain happiness, would it help us to study depression and anxiety? An understanding of depression will certainly help us in gaining insights and ideas on how to become happy, but it is more crucial to understand what contributes to joyful experiences to attain happiness. The basic challenge of this book is to discover and understand our competences to act safely.

The first angle in approaching this question is to analyze how our ancestors managed to survive in quite dangerous environments. The most famous environments in the history of mankind are the African savannas at the time the Sahara was still green. Our ancestors had to survive amid wild animals that were all hungry, and amid competing tribes who where also interested in the same food from time to time. Without their ability to act safely, we wouldn't be here now. All those who weren't able to survive the dangers of that time were kicked out of the gene pool and lost their race in passing on their DNA. The main reason why we are here is that we inherited competences to manage our personal safety. In this book, we will discuss these inherited structures.

The dangers of 10,000 years ago have now disappeared, but new dangers have arisen at such a fast pace that we have not been able to adjust our systems to them. From an evolutionary perspective, 10,000 years is rather short for DNA to adjust. Only minor changes can be achieved in such a period. So each of us starts life with an inherited safety management system that has proven its reliability but that is outdated. That system once worked well but it has to be adjusted to the new environment of the 21st century. That's probably the reason why people deny stark facts and behave unsafely although they know that they can be hurt or even killed. In most cases, the possible benefits of hazardous behavior lie in gaining a few seconds. There seems to be no rationale behind this behavior. In this book, we will discover that there actually is a rationale behind human behavior, but this rationale is often disturbed by nonconscious biases that influence our lives and our safety management (Dobelli, 2011). So if we want to improve safety, we not only have to understand our inherited safety systems and the processes involved to update them, but also the irrational biases that sometimes make us behave so terribly unsafely.

Unsafe behavior is often caused by biases in our systems.

The next question is: How can we update our inherited safety system so that we can survive in this modern world? Once we understand why people take risks and how they anticipate danger, it is possible to define a new strategy for organizing work in a safer way. Many professionals have already constructed hypotheses about this tendency to behave unsafely, and this has resulted in efficient policies and an enormous decline of unsafe behavior within companies during the last 65 years. Nevertheless, a small amount of accidents keep on occurring, and the question now is in which fields of science can we find new insights that can help us to establish a (final) breakthrough in our safety approach.

If we have to select one top source of new knowledge about human behavior that can help us in this challenge, it is the study of neuropsychology. Worldwide, almost 10,000 researchers put people in MRI scanners daily, while doing all kinds of tests with them. This results in an enormous stream of new facts about human information processing. For this reason, the first 10 years of the 21st century were heralded as the age of neuropsychology, as this field is delivering the highest number of new insights into human behavior. Now we must translate all of this new knowledge into advice for organizational behavior in general and safety behavior in particular.

Research of the brain now generates most of the innovative concepts that help us understand human behavior.

This book also takes up the challenge of translating some of the recent scientific insights in the area of neuropsychology into the modern practice of safety management, and generates new ideas on how to make the world safer. Most of the knowledge used is brain based. For this reason, the theory is named "brain-based safety."

The third challenge of this book is to define ways to externally influence people in such a way that they actually behave more safely. Employers want to guide their employees to improve safety behavior. But what kind of activities will actually influence the safety behavior of others? Answers to this question can be found through individual one-on-one coaching, in adjusting group behavior, and in organizational approaches like safety campaigns.

PERSPECTIVE ON HUMAN BEHAVIOR

There is a strong analogy between the structure of this book and the basic ideas about human behavior used in this book. The main scientific discussion about human behavior in the previous century has been the so-called nature–nurture discussion. Some believe that our genetic background (nature) mostly defines the way we behave today; others state that our personal learning history (nurture, the way we are brought up) is mostly accountable for our behavior. In the model used in this book, both nature and nurture are relevant but are still insufficient to explain human behavior concerning safety. A third element, the influence of the here and now environment, is needed to understand why people behave the way they do.

So if we want to understand human behavior in general and safety behavior in particular, we need to understand the relative influence of three interacting components:

- **Nature** refers to our roots, the result of billions of years of evolution, which are concretized in strings of DNA that form the blueprint of our physical being. These roots define the way our information processes work. They create our "hardware" that allows us to sense dangers and to act safely. At the moment we are born, our safety system is ready but almost empty of information. At that time, the highways of our brain are constructed but hardly anybody is driving on them.

- **Nurture** refers to our learning history, the content with which we gradually fill our systems so that they can actually make informed decisions. These learning processes are active throughout our life so that we can gain new insights and say goodbye to old ones. Learning processes generate our "software," which is essential for letting our hardware work. Every safety behavior is a result of a previous learning process. This learning results in physical changes in the brain, much in the same way as we see results in our muscles from exercise. The brain contains billions of brain cells that can communicate with each other. Learning creates new connections between these cells, changing existing connections or deleting them again. All learning is physically materialized in these connections.
- Our **actual environment** consists of the demands (tasks or assignments) we have to face, the social environment (team) in which we are acting, and special stimuli (information) that influence us. The environment has a direct influence on whether or not we use our inherited and learned safety behavior. Stating it differently, although a person has the perfect structure to act safely and has been trained to perform the safest behavior, he still can act unsafely due to external influences. Safety is a form of dynamic behavior that depends on circumstances.

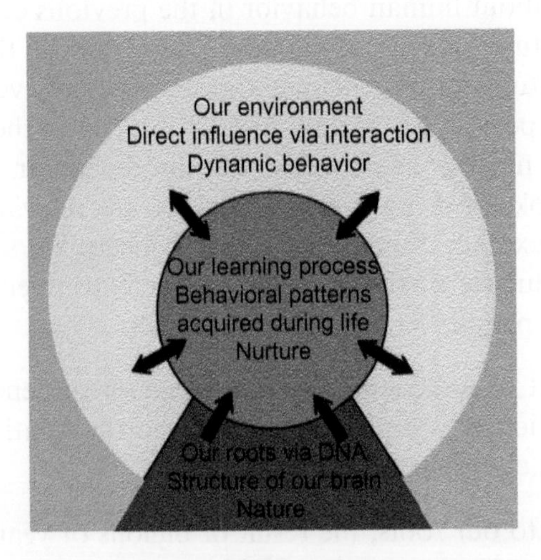

So, if we want to influence the behavior of people so that they act more safely, we need to understand our roots because they show us both the possibilities and the impossibilities. Next, we need to know how learning processes work and how we can optimize them. Finally, we need to understand the possible effects of our environment so that we can create conditions that optimize safety behavior.

THE STRUCTURE OF THIS BOOK

The following diagram represents the content of the book. At the bottom are two elements that can be seen as the roots of this approach. Together, they form Part 1 of this book. We will see that the recent history of safety management (since 1945) has impacted how we approach safety issues today. We will also analyze our roots as mankind and define what we have inherited from our ancestors. Together, this will help us to define a starting position, both as individuals and as contributors to the safety profession. The main message is that safety and survival were crucial elements in the development and transferal of our DNA, the encrypted code and master plan of our being.

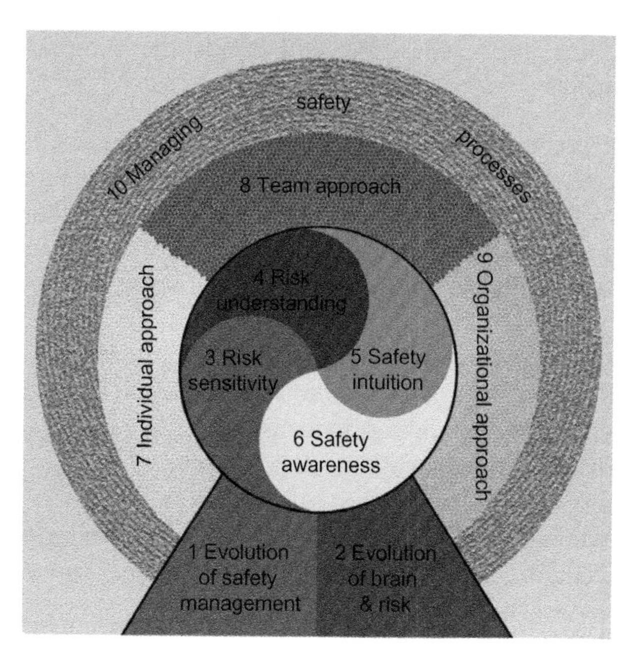

In Part 2, we will focus on our software, on how we can use the possibilities of our brain for living and working safely. In the center of the diagram is a circle. It contains four intertwining aspects that together allow us to address most of the risks in this world. Although we have inherited a good safety system, it is unfortunately not inherently programmed for the actual work and domestic environment. We have to reprogram it. Out of the four different brain functions, two are involved in detecting risks and two in handling them. The four principles are:

- Risk sensitivity—becoming sensitive to risky indicators in our environment;
- Risk understanding—understanding work processes so that we can anticipate the possible risks involved;
- Safety intuition—our nonconscious system that takes care of our basic safety; and
- Safety awareness—creating the optimum attention for the here and now work process so that we are in control of the situation, and finding new solutions for unexpected threats.

The third part focuses on the outside world and the way that this world influences us. It is visualized as a circle surrounding us and containing three approaches. We can divide the external influences into three aspects:

- The first aspect is how we can influence a person on a one-to-one basis. We will discuss the benefits of a safety coach, called the safety buddy.
- Next we will focus on how we can influence teams. This is a more cultural approach. To understand culture, we need to analyze how the behavior of people melts together in a collective repertoire called culture.
- Finally, we will discuss how an organization can influence people and how management can contribute to this process. This is called priming, influencing by activating nonconscious safety processes. Priming evokes behavior via associations. A popular form of priming is using safety posters to promote more safe behavior, but there is a whole world of undiscovered potential. Once you learn more about priming, you will recognize the principles in almost all advertisements.

A second part of this aspect focuses on the question of when and how we use these activated safety skills. A basic assumption of the brain-based safety approach is that our safety competences can be either activated or put aside, depending on what is happening in the outside world. In other words, safety competences are dynamic; even if we have them, there are circumstances in which we don't use them. Safety-minded people could be tempted to act unsafely.

The fourth and final part addresses the ways of monitoring and managing processes in organizations that should lead to maintaining and improving safety. A first aspect is how we develop a safety dashboard, a balanced scorecard of relevant safety figures. Combining these figures leads to a Safety Index, a figure reflecting the actual safety situation in parts of and in the whole organization. We will also reflect on the sustainability of trends in safety behavior and the biases that can lead to incorrect conclusions. We end with a short reflection on the impact of human resources (HR) policy on safety, especially the role of financial bonuses.

How to Read This Book

This book is the first of its kind in the safety literature. It adds a completely new view to an existing profession. From research on perception, we know that it is hard to recognize and store information that is unfamiliar. It takes some time before words and concepts gain meaning. To help readers, the information in this book is organized stepwise, from very broad and superficial to detailed and in-depth and then back to broad again. Allow yourself to digest the material. Don't hurry to finish the book.

In the book, we will reference two cases to which we will return in every chapter.

Case 1 will be introduced in the next pages. This case did not really happen in the way it is described. It is a fictitious construction based on elements of several real cases, but it could actually have happened the way it is described.

Case 2 uses a very familiar behavior: learning how to drive a car and using it daily.

As mentioned, the content of this book is partly based on brain psychology and brain biology. Most readers will not be eager to know exactly what happens beneath our hat, in the skull. They are satisfied with simply knowing the described processes. Others might feel an urge to check some of the original material. To serve both groups, most references to the brain are concentrated at the end of each chapter. Readers can skip these paragraphs without losing basic information for understanding the content.

Most chapters end with "Tips for Transfer"—questions that will help readers digest the material, practical hints, and small assignments. If readers are in the position to discuss these items in small groups, it will certainly help readers transfer the presented material into a new safety approach.

Case 1: Cooler Maintenance

Case 1 is situated in a chemical plant. Two employees (mechanics) of a contractor are assigned to do regular maintenance work on a cooler. The cooler is part of a process steam generator, which produces steam for the factory. The cooler normally is filled with oil of 200° Celsius (400° Fahrenheit) and a pressure of 5 bar. When the oil cools down, the water heats up and becomes steam.

The factory has 16 of these process steam generators; 14 are used permanently, 2 can be made free for maintenance. Part of the maintenance is to replace the pressure safety valve.

The two mechanics are familiar with this task. Once a month, they inspect two coolers. For this assignment, they go to the control room and meet the shift leader. He checks and signs the work permits of the mechanics and calls for the day shift operator to accompany the mechanics to the coolers. After a few minutes, they find out that the operator who usually accompanies the mechanics has a day off. A new colleague will replace him. The mechanics introduce themselves to the operator. Before leaving the control room, the operator checks whether the normal procedure of premaintenance has been followed, which means that the two coolers are disconnected on both sides, the pressure is released, the oil has been removed, a cleaning agent has been used to remove residues of oil, and the internal chamber is filled with nitrogen. He gets a confirmation, and they leave the control room.

Together, the three go by car to the coolers. From the parking spot, they have a one-minute walk to cooler PSG-45D. When arriving at the side platform on the first floor, the operator realizes that he has left the papers in the car. The mechanics hesitate for a moment; they should have the papers there, but nevertheless, they don't see a problem in starting. They know the procedure by heart, and they have some spare Last Minute Risk Assessment cards. They answer the 10 safety questions, fill in the card, and hand it over to the operator when he returns.

Before starting, they do an extra check, pulling a handle on the valve to manually check whether there is still pressure in the cooler. Nothing comes out of the valve, so there is no pressure difference between inside and outside. As an extra safety procedure, the mechanic closes a manual circuit breaker in the input channel. As a sort of routine, the mechanic once more checks the pressure via the manual handle.

He starts dismantling the valve while his colleague walks to the second cooler to prepare the work there. Once he takes off the valve, a small amount of gas and some fluid comes out. After a few seconds, hot oil suddenly starts escaping, which frightens the mechanic. His first reaction is to reinstall the valve. The pressure of the oil increases and while pressing the valve on the pipe, the hot oil squirts in all directions. The mechanic realizes that the pressure has become too high for him to reinstall the valve, drops it on the floor, and runs away. The operator sees what's happening and shouts to the other mechanic to leave the area immediately. The operator runs to the control room and asks for assistance.

The team manages to block the cooler, and the pressure decreases. The mechanic and the team go back to reinstall the valve and check the damage. Oil is all over the place. Due to an increased risk of fire, the area is cleared by sounding an alarm.

The mechanic then feels a burning pain on his wrist. While trying to reinstall the valve, some hot oil passed through a gap between the sleeve of his overall and glove. He consults a medical doctor who treats him for the lowest degree of burn. The 10 m^3 (350 ft^3) of oil that squirted around and dripped through the holes on the side platform created huge environmental pollution. Although the medical damage was minor, the people involved were very shocked. The costs to clean the area were significant. The incident was classified as a Class B.

Case 2 is about an experience most of us can remember: learning how to drive a car and actually driving it. In this example, we view the process of driving a car from different angles. It helps in understanding both the theory of this book and why so many unnecessary accidents happen on the road.

PART 1

Safety in Perspective

Part 1 is about our history, which has formed a mind-set we use to reflect on safety issues. We start with a retrospective analysis of 65 years of safety management, which gives us a sense of direction about how we can develop this profession to realize complete safety. This direction emphasizes the importance of the human factor in present safety management. The human factor journey began long before we were born. We inherited some of the results of the journey in our DNA.

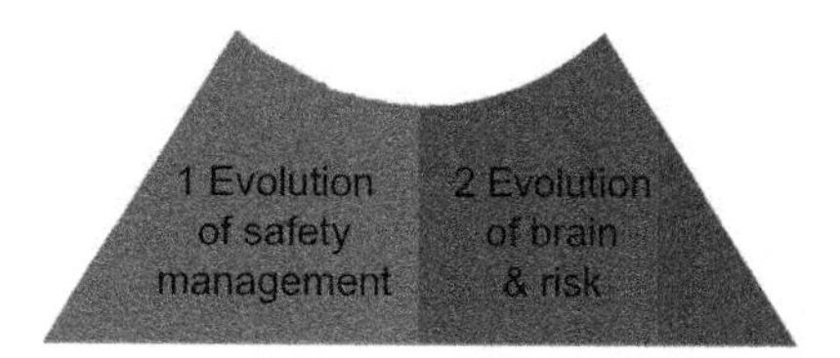

In Chapter 1, we will list the achievements realized so far in the area of safety management. We will focus on how the industry has developed a safety management system and gained competences that attributed to making work safer. Every step started with a revolution in thinking, a paradigm shift. This revolution later stabilized, again leading to a further evolution of the profession. We will look at what the next step should be.

In Chapter 2, we will look at mankind from a historic perspective and the journey from where it all began, from way back in evolution to our present society. By understanding how our history as a human race has formed our foundations and safety systems, we can also better understand our present behavior. Our brain has been developed over a time span of 300 million years. To cover that time in this book, we have distinguished five stages in evolution, each of which has contributed to how the human brain is organized today and how we deal with risk. In the first chapter, we examine how the constant battle for survival resulted in a brain that is potentially very good at managing risk, but that handles it with too much ease from time to time.

CHAPTER 1

Evolution of Safety Management

Safety management as we know it today started just after World War II. Before that period, the most common practice was to pray to Saint Barbara, protector of miners and workers. Unfortunately, Saint Barbara was not very effective. At that time, even small charcoal mines had a few casualties each year. Saint Barbara chapels were used as morgues, and many miners frequently went there to bid farewell to their former colleagues.

1.1 SAFETY MANAGEMENT LEVEL 1

Just after World War II, the chemical industry started to develop itself on a broad scale, mostly as an offspring of other industrial activities. As technology improved, the risks involved with the production process increased. One example is the experimenting with carbon–fluorine connections by DuPont, which eventually led to the discovery of Teflon. Because this compound was highly explosive, the company realized that production could only be started if it took extra safety measures. Without these extra measures, the factory would explode even before production really started. At the same time, the world was recovering from the effects of the Second World War. So many men were killed to regain freedom. In this light, it became unacceptable to place more men in danger just because of work. From today's perspective, it is an obvious decision (at least in some regions of the world), but at that time, it was a real breakthrough in thinking about labor: Work should be possible without fatal accidents. Factories were developed much more safely, equipment was tested for safety, and all workers started to wear safety protection adjusted for each task. This breakthrough can be qualified as a paradigm shift. The new paradigm was that each employee should reach his retirement alive—overlooking the fact that silicosis for miners was still widespread. In the paradigm shift, people concluded that human life was more important than the profit of the company. It would take us many years to find out that most safety costs showed a positive return on investment due to a more stable process performance.

Paradigm 1: Work should be possible without fatalities.

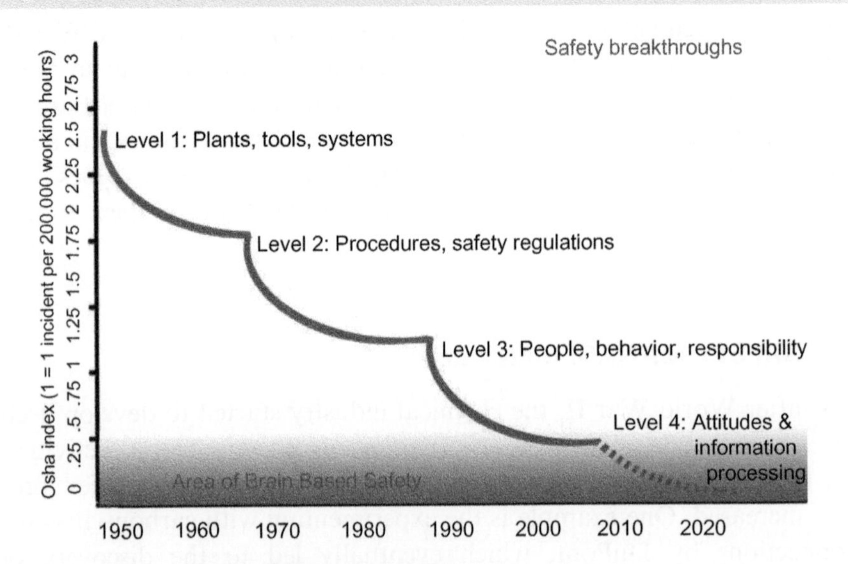

1.2 SAFETY MANAGEMENT LEVEL 2

The new way of thinking and working proved effective, as the casualty rate dropped dramatically. In the 1960s, the incident rate in the chemical industry stabilized at a certain level, and many companies around the world embraced the idea that more was needed to create a safer work environment. Then the focus shifted to those using the equipment: the employees. They still could act unsafely in safe environments. Controlling the behavior of employees would improve safety levels. With this in mind, a second breakthrough was the introduction of procedures and protocols to regulate work and work behavior. Roles and responsibilities were defined, tasks were described, and employees were instructed to work according to procedures. Certification of people, parts, and procedures became standard. Technical risk assessment was introduced. The paradigm shift was that a safe plant alone was no guarantee that accidents would not happen; instead, teaching employees how to perform a certain task would increase the safety. Focus shifted to acting according to the book. Sticking to rules is an important way of regulating behavior in many regular activities today. The downside is that it only works in standard situations.

Paradigm 2: Regulating behavior via rules will reduce the amount of safety incidents.

1.3 SAFETY MANAGEMENT LEVEL 3

In the 1990s, companies were again facing the fact that the incident rate stabilized at an unacceptable level, for that moment. More of the same interventions no longer reduced the incident rate. A new breakthrough in approach was needed. Not all the behavior in the workplace could be described in procedures. Besides that, people don't like to be prescribed how to behave; they have a natural tendency to break rules that don't suit them. Procedures require constant management attention that cannot always be provided. To address this, employers needed to find another mechanism to control human behavior, and the answer was found in looking at the social environment. Thus, the focus shifted to personal behavior. The paradigm shift was that prescriptions on how to carry out tasks were not sufficient for generating safety. Employees had to be convinced to follow behavioral guidelines under all circumstances, not only when prescribed; personal accountability became part of the safety policy. A second element in this renewal was the focus on social control in addition to management attention. It was recognized that the behavior of the team had at least the same influencing effect as the supervision of the boss. By recognizing that, safety management also became a cultural issue. The new approach embraced a social view on behavior and behavioral change. The team was seen as the most appropriate level to enhance and guard safety behavior. The ideal was that employees mutually challenge and stimulate each other to embrace proactive ways of behavior: solving an emerging problem before it actually became one. Theories like the High Reliability Organization (HRO) and Hearts and Minds operate successfully on this level.

Paradigm 3: Safety will increase if people take personal responsibility for their own and others' safety.

This breakthrough again helped people reduce the number of incidents, but the incident rate has stabilized at a certain level of OSHAs[1] during the last few years. Among some, there is a belief that, although it is not easy, work free of incidents is possible. Strangely enough, we now have the same discussion as those at the onset of safety management, only the subject has changed slightly. At that time, those who thought that work without fatalities would not be possible challenged those who had the ideal of a workplace free of them. The only difference in the discussion is that we now talk about incidents instead of fatalities. Fortunately, the idea that all people can leave their work in the same health as they began is winning support. As in every breakthrough in thinking, a solution that is the result of more of the same approach does not work. We need a new, fourth paradigm shift to create brain-based safety.

1.4 SAFETY MANAGEMENT LEVEL 4

The crucial question now is how do we create this new breakthrough in management thinking? A problem with a paradigm shift is that you don't see it until you have made it. Fortunately, the history of safety management can help us.

When we analyze the first breakthrough, we see that safety started with a technological approach focused on building safe systems (for example, equipment or logistics). Engineers developed the roots of safety management; even today, we can see that the majority of the

[1]OSHA is the abbreviation of Occupational Safety and Health Administration, an American institute founded after the OSH Act. These days, the expression OSHA is used to refer to a recorded safety incident and an indicator of the number of incidents per career per employee. OSHAs can be weighted in categories, varying from fatality to first aid. Unfortunately, there is no generally accepted standard OSHA index. In this book, the OSHA index of 1 indicates that every employee or contractor has the probability of being the victim of one incident during 200,000 hours of work. On a site of 1,500 employees and contractors, this would be the equivalent of 13 incidents per year.

safety professionals have an engineering background. The second breakthrough was focused on how to use the systems: the procedures and protocols. Man is seen as a robot that needs to be programmed to follow the rules that safety specialists have developed and described. The third breakthrough focused on the users and their community: the responsibility to act safely, especially when it is not possible to describe the work to be done in procedures. One is asked to act according the spirit of the rules and to stimulate each other to do the same.

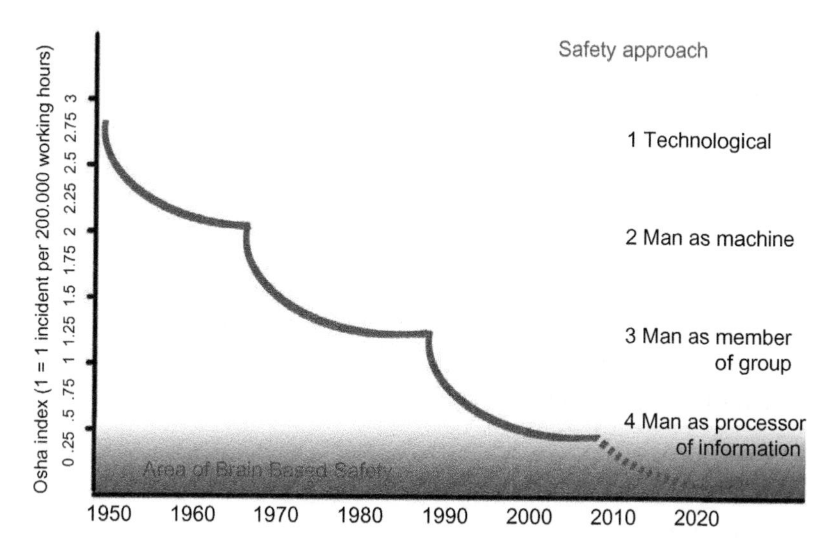

The trend is clear: Further progress in safety management can only be achieved by focusing on the human factor. In other words, the final breakthrough to complete safety can only be found deep inside the actor: at the moment a person perceives and estimates the amount of risk involved in a certain activity and at the point where he chooses his actions, where he has the option to act more or less safely.

Paradigm 4: Safety will increase if we can understand and influence the human factor in safety behavior.

History teaches us that safety management has a very consistent value system (work should be possible without harming people), in which the norms are upgraded from time to time. Safety management started from a technological approach outside the human system and moved step by step via the expertise area of labor psychology, via the more social team approach into the field of intrapersonal psychology. If we are able to influence the mind of a person at the moment she decides what to do and how to act, it will be possible to generate safer work behavior. We are in need of new areas of expertise that can teach us more about human behavior in general and the process of estimating risks in particular.

Fortunately, recent developments in neurobiology and neuropsychology are deepening our insights into how people make decisions and what kind of brain processes are involved. One of the major problems is that decision processes in the brain are far from conscious and rational. In these processes, different sources of "information" are used, from very logical analyses to very intuitive emotional estimations. The main challenge is to influence the internal processes of weighing risks. The holy grail of safety management can be found by understanding the essence of human behavior, the principles by which we perceive our environment, digest the information, and come to action. The challenge of this book is to transfer all the recent knowledge in this field into understanding safe behavior and into ways of influencing these processes.

1.5 SUMMARY

Safety management started as an economic necessity and as a result of changing values. During the last 60 years, the developments in this professional area can be described as a combination of revolution and evolution. Each revolution in approach is accompanied by a paradigm shift, a new perspective. Stepwise ideas have changed from the notion that work should be possible without casualties, via the notion that safety can only be accomplished if employees stick to certain rules and contribute to safety as a result of personal responsibility, to the notion that a collective approach can make it possible that everybody leaves work in the same health in which they started. Each step contributes to an integral safety approach and enlarges the focus. Safety management has changed from a pure technical approach to a multidisciplinary one, integrating engineering with labor and personal psychology. The final step is to understand human safety behavior and to influence that from all possible perspectives. The recent developments in neuropsychology can attribute to this understanding.

TIPS FOR TRANSFER

Tip 1: Assess the Present State of Safety Management in your Organization

As far as you can assess the quality of the actual current safety management, how mature is the state of the art in your organization on each of the following four levels? Explain your answers using some keywords.

Level 1: Plants, tools, systems

Level 2: Procedures, protocols, safety regulations

Level 3: Culture, behavior, responsibility

Level 4: Attitudes, beliefs, information processing

Tip 2: Use the Checklist to Make a Diagram

Score each of the following questions on a 5-point scale, varying from "I don't agree" (score 1) to "I fully agree" (score 5). Add the totals per level (minimum 4 and maximum 20) and draw a diagram.

Level 1: Plants, tools, systems

1. The technical installations in my company are safe.
2. The maintenance of our installations guarantees maximal safety.
3. Employees have and use all tools needed to act safely.
4. Regular safety inspections and deviations are addressed.

Level 2: Procedures, protocols, safety regulations

5. All external regulations (for example, government policies) are integrated into the company's way of working.
6. Employees know all rules involved in doing their specific tasks safely.
7. Employees understand the reason behind each of the procedures.
8. If a person or a team neglects safety rules, they receive feedback to change behavior.

Level 3: Culture, behavior, responsibility

9. Employees are dedicated to work safely.

10. Even if management needs to make painful decisions, safety is always guaranteed.
11. Like employees, contractors are instructed in safe working.
12. If somebody behaves unsafely, he is corrected by his peers.

Level 4: Attitudes, beliefs, information processing

13. Employees are instructed to sense all risks involved while working.
14. Employees understand their own work processes and possible dangers involved.
15. People who work here have a true conviction in behaving safely.
16. Safety at home is treated with the same care as safety at work.

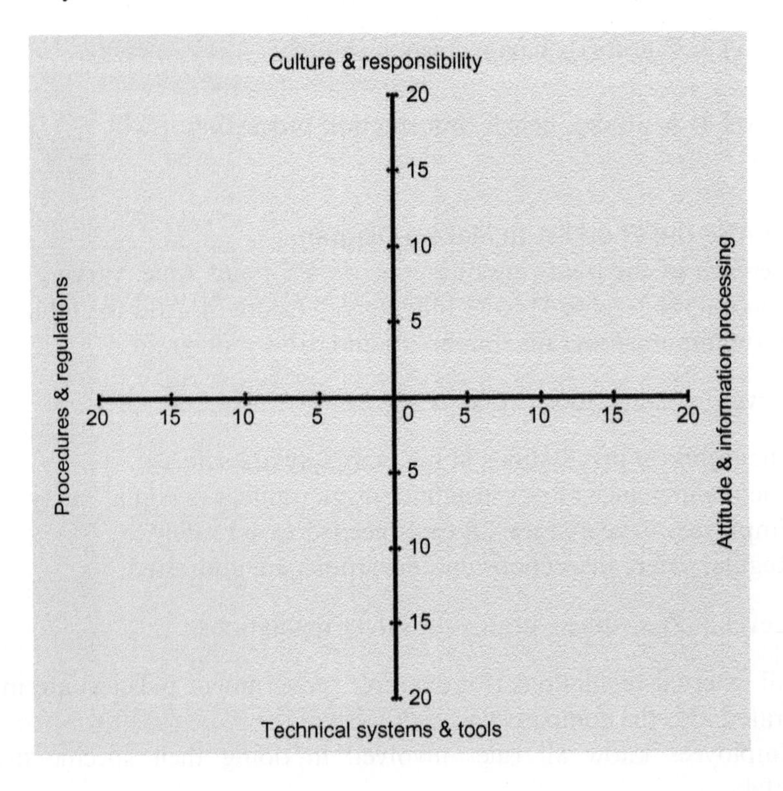

CHAPTER 2

Evolution of Brain and Risk

If we acknowledge that we can renew safety management only by addressing the human factor, we need a deeper understanding about us as humans. We have to realize that we don't start from scratch when we are born. The roots of today's behavior lie in evolution of mankind. Thousands of generations have experimented with all kinds of safety behavior, and only our most effective ancestors reproduced. During evolution, a blueprint developed that offers us both possibilities and constraints for behaving safely. Understanding this blueprint is key to understanding human behavior.

Biologists state that human evolution has only been possible due to a combination of personal survival and reproduction. To enhance these two, man has evolved as a group animal. Living in a group diminishes external dangers and gives better opportunities for raising children. So survival, reproduction, and group behavior are key elements in our existence, and our brain is developed in such a way that it can ensure these three basic functions. When we zoom in on personal survival, we can see that we survive due to a combination of physical integrity and good nutrition. Safety behavior guarantees that the body stays in good shape. Managing serious risks is a natural and essential element.

To gain a deeper understanding of our present safety behavior, we can divide evolution into five stages. We will now tour through each of these stages, starting 300 million years ago and ending today. The most relevant items of our genetic blueprint will be discussed.

2.1 STAGE 1, FROM 300 MILLION TO 200 MILLION YEARS AGO—THE DEVELOPMENT OF THE BASIC BRAIN

The first stage of the tour starts roughly 300 million years ago, the period in which the first reptiles—like lizards and snakes—developed. The reptiles already had a brain system with basic functions we can still distinguish today in the human brain. One part of our present brain still has a strong resemblance with the brains of reptiles. This part is called the "basic brain," or sometimes even the "reptile brain." The basic brain takes care of all our basic functions for survival: being awake, sleeping, breathing, eating and drinking, sex, and basic coping behaviors like fighting, fleeing, or freezing. All our sensory information passes for a first stop in this basic part of the brain. During this first stop, all the information is quickly processed and subjected to a first scan. This scan is on a nonconscious level and focuses on the three basic missions in life: survival, reproduction, and cooperation. A safety check is one of the important elements of survival and part of the first stage of all our perception. The sensed information is then sent to other areas of the brain, where the conscious seeing, hearing, smelling, and touching is done. A direct result of the nonconscious prescanning can be that the attention of the eyes and the ears is focused on particular stimuli. This focusing is done to collect more and better information. So we can focus our eyes on the floor just to check for a potential risk (for example, something that might be a snake), on the behavior of others (for example, to see where their attention is focused or whether they can be trusted), or for attractive options for creating offspring (neuropsychology confirms some of Freud's statements; on a nonconscious level, we are always preoccupied with sex). This first scanning is finished after 0.1 second. It takes another 0.4 second to consciously notice these new stimuli and to react on them. In the meantime, the basic brain can prepare a possible action (raising the blood pressure, increasing the heart rate). This helps us to be more prepared just in case something dangerous is happening. We conduct this nonconscious scan—with a total throughput time for conscious recognition of 0.5 second, which can be too slow in case of an emergency—to help fully prepare the body to act once we consciously notice a serious threat.

The basic brain works on a nonconscious level and regulates our basic survival functions, hardly giving us any cues.

We can recognize this 0.4-second delay when we throw garbage in the trash but then realize the trash has been emptied but a new liner has not been placed in the can. When we open the trash can and automatically start throwing in the garbage, we notice that the liner is not there, but we cannot stop the movement of throwing in the trash, because it takes another 0.4 second to stop the movement. Within this time span, the garbage already lands in the can.

Our conscious system always lags a half second behind the actual facts.

As the basic brain acts on a nonconscious level, all we can sense from these activities is a very basic and subtle feeling, positive or negative, without knowing why. We will explore this point later in Chapter 5, when we discuss the safety intuition. General feelings like pain, happiness, unrest, or panic originate in this system. The impact of the negative impulses is larger then the positive ones. The evolutionary rule is that it is easier to recover from a false alarm than from an unexpected threat. Pain is the first signal that something is going wrong. The basic brain has developed a pain system with separate information channels. This system starts with pain sensors in the skin that send signals via the spine to the brain. This pain system is very complex and is still not fully understood. It is slower then the other senses, like touch or temperature. So we first feel that something is very hot, and a half second later, we experience the pain that is involved with burned skin. This probably has to do with the original fleeing behavior: Run first, check the damage later. We may feel pain, but we don't know yet how damaging the situation is. If a knife touches a finger, we first have to look to determine if the finger is really hurt. If we actually have a wound, the pain signal slowly becomes stronger, just to take care that we allow the wound to recover.

Pain signals are the most original signs of external threats.

Negative pain feeling might be a sign of a life-threatening event; we never have a negative pain feeling when the system experiences something positive (for example, tasty food). To survive, we have learned to focus first on the negative and then to try to avoid this. This principle developed in what we call "loss aversion," which will also be discussed in Chapter 5. We can recognize this principle very clearly when we have an appraisal talk with our boss. If he mentions five positive points and one negative point, we will probably give more attention to the negative point than the five positive points, even if this is introduced as a point for approval.

First avoid the negative, than experience the positive: the origin of loss aversion

All brain parts usually work together, but in the case of real danger, the basic brain has the opportunity to act solely. This acting is primitive and impulsive. The male basic brain has three options, usually referred to as the three Fs: fight, freeze, and flight. Fighting means attacking and eliminating the danger. Freezing means just acting as if you are dead, a very effective way to neutralize an aggressive environment. Fleeing means running as fast as you can to try to beat the danger. The female basic brain has two other basic options, sometimes summarized as "share and care." Sharing means that females have the tendency to start communicating to collectively counter the problem. Caring means strongly focusing on the environment, taking care that everybody is all right. We have to realize that each of these patterns once belonged to the best practices of safety management. The gender distinction probably has to do with the different roles in the past: Females caring for a lot of children had to make a collective plan for themselves and their social environment in case of a threat, whereas males experienced their most stressful moments while hunting.

The basic brain is also involved in a final check of our behavior just before we actually execute it. It has the possibility to block this behavior in cases where something might go wrong. In terms of risk management, the basic brain gives very basic and hardly conscious cues that something is not right. We can feel uncomfortable and just have a vague idea of why. Although the basic brain works on a nonconscious level, it can be trained to focus on certain objects or situations. All fear conditioning for hazardous objects starts here. The lesson to be learned from this section is that influencing these nonconscious processes can enhance safety management. In Chapter 3, we will discuss ways to become more sensitive to dangerous stimuli.

The basic brain has no direct access to our conscious system and influences us via indirect cues like a feeling of unrest or a strange feeling in our guts.

Last but not least, we have to realize that the basic brain operates in direct contact with the actual here and now situation. In the most recent part of the evolution, man has become the best planner of all animals, but we still have biases in our system due to the overappreciation of the actual situation. Kahneman (2011) talks about the "what you see is all there is" bias. When we need to act safely, we sometimes neglect the bigger picture and base our action on the actual stimuli.

Summarizing, the basic brain is the first support for acting safely. It works fully on a nonconscious level and prescans the world for possible dangers. Although our conscious perception is always lagging 0.5 second, the nonconscious perception prepares us for possible dangers. In case it detects such a danger, it will direct the attention to the stimulus and prepare the body so that it is ready for possible action (for instance, in fright from an unfamiliar sound, the heart rate increases and the sugar level in the blood rises). Pain is the first signal of danger. Due to the basic brain, we give more attention to negative experiences compared to positive ones. Although we cannot directly experience the output of the basic brain, we experience hunches that tell us what to avoid.

2.2 STAGE 2, FROM 200 MILLION TO 2.5 MILLION YEARS AGO—THE DEVELOPMENT OF THE EMOTIONAL BRAIN

The next stage in evolution started approximately 200 million years ago, with the appearance of mammals. Mammals take care of their young after birth; they all have at least a temporary relationship with their offspring in the period that they feed them. Some mammals—like humans—are group animals; they survive by joining together, and they raise their offspring within a community. Group living means that we have to manage relationships. To regulate relationships, the brain has developed a second layer called the emotional brain or the limbic system. The emotional system manages how we personally relate to the outside world.

Here again, we recognize two basic directions, positive and negative, that appear to us in the emotions of love and anxiety. We have some innate anxieties, for example, for extreme weather conditions, insects, or snakes. Besides that, we are programmed to stick together. As group animals, we need a warning system for when we are losing connection with the group. This warning comes in the shape of anxiety of loneliness and social exclusion. All other anxieties are learned. Anxiety can be seen as an anticipating emotion to prevent pain. Experiencing pain leads to avoiding anxiety. We will see later in this book that safety management starts with creating associations between objects or situations and anxiety. Without this conditioning process, there is no basis for safety behavior.

Anxiety is anticipating possible pain. It is a basic ingredient of safety behavior.

The brain keeps a record of all learned safety experiences, the associations of anxiety with objects or situations. We store information on, for example, who to trust or distrust, which situations might be threatening, and which objects should be avoided. All pain- and anxiety-related experiences are stored in a special area in the center of our emotional brain. If this area is damaged—for example, due to a stroke—surviving is really difficult. We simply don't experience dangers anymore. Although we like to see ourselves as rational and logical, this emotional database has a huge impact on our decision-making process. If we transfer this knowledge into safety management, the emotional brain gives us very specific clues about risky elements in our environment, as well as people, objects, or situations. In each clue, a minor touch of anxiety is included. Without feelings of anxiety and pain, we cannot survive.[1]

Mammals learn individually but also as a group. Emotions are shared, and this enhances the learning process in the group. Signs from group members that something might be threatening are very important for the others. The group will immediately act, for example, in a flight reaction, when one of the members senses danger: the herd instinct. We have learned that the chances of survival increase when

[1]There are people who are born without the ability to experience pain (analgesia). They usually die before they are 20 years old. Children with this condition often suffer oral cavity damage (such as having bitten off the tip of their tongue) or fractures to bones. They also have difficulty sensing anxiety.

you follow the fleeing herd compared to making up your own mind first and checking whether there really is a danger. Historically, most independent actors who didn't flee with the group at the moment of actual danger didn't survive that danger and had no offspring. Even today, we can see that this herd instinct is active in keeping each other out of harm's way. But the herd instinct has strange side effects. Due to this instinct, modest swings in our economy develop into huge fluctuations on the stock market. We want to maintain our safety in the group, so we sell our stocks when all others are selling their stocks, even if the prices are low, and we buy when everybody else is buying, even when the prices are too high.

A bank run is a perfect example of our herd instinct.

Strangely enough, the herd instinct can even have an opposite and damaging effect. Although it contributes to managing safety as a group, it can lead to collective trespassing of safety rules. The psychological impact of a group can be so strong that even an independent and safe position can be experienced as threatening, especially if you are the only one in that position. Social exclusion directly activates the pain center, so we want to stick to the group, even if the group is exposing itself to dangers. If we are abroad in a tourist environment and have to choose between two restaurants, we will probably choose the one with the most guests inside, although we will have to wait longer before our food is served. We just assume that the other tourists have a better knowledge of the quality of the restaurant and we will follow them blindly. So, influencing teams is one of the main ways to increase safety within companies.

A team has a strong impact on safety behavior.

It was in this period that our basic relational patterns developed—patterns that we still have today and that can be used in developing safety behavior. Most human herds, tribes, were between 30 to 50 people. The tribe was the basic unit in which survival, reproduction, and cooperation were safeguarded. Tribes could cooperate in case of a mutual external enemy or compete in case of food scarcity. Each tribe was divided into several families with approximately ten people and three generations. The power structure was simple. A tribe had a tribe leader who deserved his position because he was strong and clever enough to lead the tribe in catching food. Each family had a head, usually the father or the grandfather if he was still alive. Among the siblings was an age-related hierarchy. Older siblings took care of the younger ones. In Part 3 of this book, we will reuse this tribe structure to create safety behavior. The roles of older sibling (the experienced and wise colleague), father (direct boss), and tribe leader (department manager) are especially important. There is no fundamental difference between raising children and managing people.

Relationships with the human herd are still active today.

Another tactic of our predecessors was to steal food from another tribe that was more successful in hunting. In times of food scarcity, other tribes or clans became enemies. Tribe members learned to make distinctions between those that belonged to the group and those that did not, the in-group and the out-group. Even today, members of different departments within the same company can become enemies when budgets have to be reduced. In our mind-set, safety is strongly related to the group of people that belongs to us and can be trusted. If we want to promote safety, the in-group is the best vehicle to do this.

The in-group is most suitable for teaching safety measures.

Summarizing, safety management has a strong social and emotional component. We create safety together and anxiety is a crucial element in all risk perception. As far as this risk perception is innate, it must be programmed for objects or situations that exist in our modern life. We have to learn the dangers of this world during our life.

2.3 STAGE 3, FROM 2.5 MILLION TO 10 THOUSAND YEARS AGO—THE DEVELOPMENT OF THE MODERN BRAIN

The third stage started approximately 2.5 million years ago. This is the period sometimes referred to as the Stone Age. The *Homo sapiens* emerged and developed in this period. We have evidence of human beings going back to 200 thousand years ago, but they are probably much older. In this period, the brain developed a third layer called the modern brain, the neocortex. This system is dedicated to perception and adds cognitions to what we perceive. Speech, music, geographical orientation, and fine motor behavior are located in this part of the brain. The modern brain helps us to understand the world. It can reason, reflect, plan, and put things in perspective. It makes us very human and allows us to overlook the fact that our genes are still 99 percent shared with chimpanzees.

Let us analyze the risk assessment made by the modern brain in more detail. In the Stone Age, the brain developed into the mysterious organ we know today. During this period, our predecessors were living as nomadic hunters in the savannas and grasslands of northeast Africa. They developed simple tools, mostly for catching and preparing food. The major risks were height, heat, adverse weather, wild animals, insects, and competing tribes. As nomads, our predecessors traveled around, visiting new areas again and again, not knowing what the dangers are. For this reason, everything that was new and unknown was treated with suspicion. The modern brain made mankind so clever that they could survive in high-risk environments though not being the strongest or fastest species. The whole neurological infrastructure to act safely and to survive under heavy risk circumstances was developed in this period. This structure is now part of the foundations of our brain.

> The modern brain hosts executive functions like reasoning and planning, and is involved in technical risk assessment.

Mankind learned to find a balance between being cautious and taking risks. Our predecessors were hunters, and one of the main tactics for getting food in that time was to steal the leftovers from lions and tigers who had just caught a big meal, for example, a zebra. This theft involved a lot of tactics, for example, determining the best moment to strike and how to escape if the plan was not working. This was a risky venture but needed to be done to survive. Lions ate those who took too many risks, and those who were too afraid to take risks starved. We can also see a gender differentiation here. The males, who were mostly active in hunting, needed to take bigger risks compared to the females who were more the gatherers of fruits and the caregivers for the children. Today, we can recognize that the anxiety system is better developed in females than in males. In most couples, the mother is more concerned and protective toward the children, and the educational style of the father is more often seen as laissez-faire. This differentiation has nothing to do with a personal lifestyle; it is inherited from our ancestors. In safety training, it is easier to make females more sensitive to dangers then men.

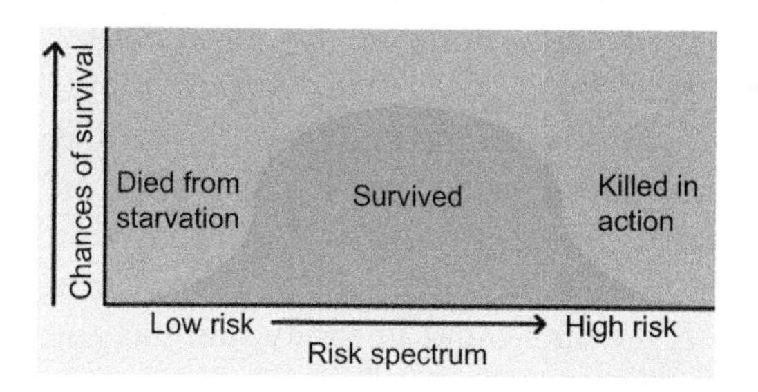

Anyway, the survivors of this period managed to find equilibrium on the risk spectrum. We have to realize that we are the descendants of a race that is used to taking risks. Acting 100 percent safely is to some extent contradictory to our nature. So if we want to reach complete safety, we will have to compensate for part of our inheritance.

The biggest risks were taken during hunting, and the largest dangers were met while being exposed to the aggression of competing tribes. In both cases, our ancestors had to survive via a battle between groups. Cooperation and support were needed to succeed together. In this period, humans were trained to keep an eye on each other and to communicate about plans and perceived dangers. It is easier to watch others acting in a dangerous environment and to understand the dangers they are in than to watch one's own position and the dangers involved in that position. It is also harder to see a loved one dying compared to seeing yourself die, so we are emotionally involved in taking care of the other. Our competence in estimating the amount of danger someone is in is much more strongly developed than the competence to estimate our own danger. Connected to this, is it also easier to estimate the safety behavior of others than our own safety behavior. These two processes are organized in a completely different way in the brain. Feedback can fill this gap between the two perspectives.

There is still another aspect that we learned during that period, which is relevant to risk management today. Hunters always had to face the fact that food supply could be very unstable. To compensate for the risk of periods of food shortage, the brain learned to consider serious risks while planning behavior and to compensate for possible failures in the plan: the origin of safety margins. If the average food consumption of 2,500 kilocalories a day is sufficient, the brain will ask for 2,800 kilocalories, in this way saving some fat (a physical safety margin) for periods in which no food is available. We can recognize similar compensating processes in other areas of life. If having three children was considered enough to safeguard survival posterity, the programming was to produce eight so that at least three of them would survive. Clever men build in safety margins to compensate for possible setbacks. In everything we do, we build in safety margins.

The brain is fully designed for avoiding risks. It regards risks as a normal aspect of life and anticipates on them.

In the example of driving a car, managing safety boundaries is one of the most crucial competences of a good driver. On the road, we meet hundreds of other drivers who all have their own driving habits. We constantly have to manage the distance between our own car and the cars around us, with a minimum depending on the speed and the circumstances of the road. The perfect distance allows us to reduce the speed of our own car safely, even if the car in front has to break quickly. As soon as you see children playing with a ball, you take into consideration that the ball might roll in your direction and that a child may run after it. Good drivers anticipate in such a way, by changing the speed and the route, that they can avoid the ball and children in all possible cases. That is the core of building in safety margins.

By building in safety margins, the modern human brain has developed itself as a long-term planner and has incorporated risks as an integral part of life. The safety margin needs to consider not only what is there here and now but also what could be there and then. Our system is not built to prevent all hazards but to find an optimum level of risk taking.

2.4 STAGE 4, FROM 10 THOUSAND TO 200 YEARS AGO—THE DEVELOPMENT OF RISK TOLERANCE

In the latest stage of evolution, the Bronze (10,000 to 7,000 years ago) and Iron Ages (from 7,000 years ago), our environment changed rapidly, and the brain was able to adjust to this only to a small extent. *Homo sapiens* became farmers and were able to produce many tools. All over the world, they created settlements in fertile areas. Suddenly, it became possible to acquire property. Competition for the same land led to wars. Within settlements, some of the previous natural dangers were present only in mild forms. The control of food production increased, and planning became more important. Man became the biggest risk. Compared to the previous periods, most of the conditioned alarm systems were hardly activated. The danger arousal and anxiety systems in the brain developed a high level of risk tolerance to the relatively few risks still present. We mostly felt unsafe when meeting fellow citizens who appeared different. We still associated them with members of different tribes that could steal food or our children. The most common stimuli didn't elicit anxiety anymore. Our warning systems had become lazy, and we became quite tolerant of possible risks in our environment. In modern safety management, we have to compensate for risk tolerance. The lesson to be learned here is that we have a tendency to stay in our comfort zone, even if this is not appropriate.

The brain has developed a tolerance against risks.

2.5 STAGE 5, THE LAST 200 YEARS—THE SUDDEN INCREASE OF NEW DANGERS

In the last stage of evolution, since the Industrial Revolution, our work and home situations have developed dramatically. In the more developed parts of the world, we do not have to fight for food anymore, and the most aggressive animals walk on a leash. The once-useful innate danger activation is seldom triggered in this modern world. At the same time, danger comes from new sources, from an environment that is artificial to the original brain. We have developed so much technological power that we can easily destroy the whole world. What has been developed in billions of years can be eliminated within a day. At work, we meet new technologies, tools, and substances that can be very dangerous. Impacts of the new technologies are enormous regarding their speed, power, and strength. At home, we have to deal with new equipment, instant boiling water, and unstable steps. On roads, we have to deal with a very dangerous traffic system that causes many casualties all over the world. Our brain has hardly any natural warning systems for these new dangers. Although we inherited a very well-functioning safety management system, the changes have been so fast that our brains need full reprogramming.

> The safety system of the brain has to be reprogrammed for new dangers.

That is the challenge for modern risk management: to use the safety structures in the brain as much as possible by reprogramming them for the new dangers. This reprogramming is part of the personal learning history of each individual. We have already seen that knowledge can easily spread amid a group, especially an in-group, but each individual still has to create a personal learning curve for acting safely. In the reprogramming, the old inherited sensitivities can be linked to new stimuli. We can call the result the second nature, because it helps us to adapt to a new world and to adjust to our required roles in society.

While reprogramming, we develop domestic and work safety behaviors. We learn that a stove can be very hot and that machinery can be very dangerous. An essential part of raising children is to teach them the dangers of this world. The same can be said about the role of employers. They have to teach their employees how to work safely, as employees cannot rely on their previous experience. Safety education is an essential part in training, pointing out potential consequences from underestimating serious dangers. To illustrate this, just look at the amount of serious accidents in the domestic environment. More than 50 percent of all emergency treatments originate from accidents in the home. Even at home, our risk tolerance is much too high, and the dangerous stimuli are not sufficiently programmed.

Employers have to teach new employees the risks involved in work just as we teach our children the dangers of the road.

2.6 CONSCIOUS AND NONCONSCIOUS

So far, we have addressed the five stages of evolution and referred several times to conscious and nonconscious behavior. Before linking our history of development to safety management, a short recap of the place of consciousness might help to clarify the importance of the nonconscious.

Somewhere during the evolution, mankind developed consciousness on top of automated behavior. It is not particularly known when this happened, but we can see that several other species, like chimpanzees and dolphins, have also developed some form of consciousness. We enjoy our consciousness, and it makes us very human. We use it for reflecting, planning, communicating, reading, learning, and responding to unknown stimuli or carrying out unfamiliar tasks. However, the majority of our activities is still nonconscious. Behavior stems mostly from automated systems in our nonconscious domain, especially routine activities like eating, walking, and common motor tasks.

The nonconscious regulates all automated behavior but can also be involved in complex mental assignments.

In the example of driving a car, everybody has experienced moments in which driving is fully nonconscious. We can even drive, for example, for 15 minutes and suddenly realize that we have traveled a substantial part of our route already. The more familiar we are with that route, the easier it becomes to drive fully nonconsciously. During these 15 minutes, we have avoided many potentially dangerous actions that could easily have gone wrong, even with fatality as a possible outcome. However, all went well; our nonconscious system often helps us safely through a day.

The nonconscious is also involved in complex tasks. Research shows that we can even finish complicated tasks more successfully on a nonconscious level, compared to a conscious level. If we ask two groups to decide which of eight apartment options is the best to rent, the group that is instructed to spend time making lists of pros and cons will do significantly worse than the group that is distracted halfway through and is asked afterward to give a first impression (Aarts & Dijksterhuis, 2002). Especially when the elements of each option are hard to compare (price per unit surface, location, view, quality of kitchen), the conscious approach scores below the nonconscious one. In other words, the nonconscious originated from doing simple tasks but contributes now to even very complex mental activities. The nonconscious is always active and the conscious can be invited to participate. Connecting this statement to safety, we can conclude that safety management is processed at a nonconscious level, and that the conscious part of our existence is sometimes invited to contribute to this safety process.

The next question is what is needed to invite the consciousness. The general procedure is that the brain first estimates whether a task can be done automatically. If this is the case, there is no need for conscious involvement and the task will be carried out fully nonconsciously. But if the nonconscious is faced with an unknown situation or problem, with no routines to guide the handling of the appropriate task, the consciousness will become involved at that moment.

The consciousness is called upon when we face unknown situations or tasks.

Case 2

In the example of driving a car, we can drive on the highway for 15 minutes without any consciousness (while listening to the radio) and "wake up" only because someone in front of us is doing an unexpected maneuver to which special attention is needed.

Dijksterhuis (2007) uses an analogy to describe the relationship between the conscious and the nonconscious. He compares consciousness with a stage in a theater. While all the actors are very busy with nonconscious tasks somewhere in the theater, a few of them are temporarily active on the stage and doing something special. Our attention is drawn to those actors on the stage, and we neglect all the others. Estimates about which percentage of our activity is conscious vary from 0.1 percent to 0.00002 percent, depending on how many processes are involved in the total sum of behavior.

If we transfer this knowledge to improving risk management, we must realize that the main task will be to influence (condition and program) the nonconscious processes in our being. Directing the conscious part of our brain to act safely will definitely help, but the main effect can be achieved by changing our nonconscious behavior. So if a new safety policy is introduced, for instance, this policy will be addressed to the conscious part of our brain. All communication will be focused on understanding and accepting this policy. Usually the project of changing a policy stops as soon as the message has been understood and accepted. That is unfortunately no guarantee that the message will be incorporated into our nonconscious patterns. The next time we have to perform, we still tend to use the old behavioral patterns that are stored in our nonconscious systems. To improve safety management, we need to do more than distribute and explain policies. In Chapters 6 and 7, we will go deeper into this.

Feeding our conscious system with a new safety regulation is no guarantee that the nonconscious will incorporate this into its behavior.

In the case of driving a car, if you have lived in the same area for a long time and are familiar with a certain road on which you are allowed to drive 45 miles (70 kilometers) an hour, you will have a hard time adjusting to a reduced speed limit of 30 miles (50 kilometers) an hour. If you are driving with your nonconscious, you will have a tendency to stick to the old pattern and to drive at the previously allowed speed.

2.7 COMBINING THE TOPIC OF CONSCIOUSNESS AND THE THREE PARTS OF THE BRAIN

It is really difficult to gain conscious access to activities in the basic brain, which is mostly active in the realm of the nonconscious. Fakirs (Hindu ascetics) claim that they can access conscious control of basic brain functions (for example, lowering the heart rate by exercising breath control) with a lot of training. We can only experience some results with these basic functions (for example, by experiencing hunches or gut feelings). When experiencing a gut feeling, the emotional brain serves as a translator of some cues in our basic system.

We can have more access to our emotional brain, although this part of our system also functions mostly on a nonconscious level. In most cases, our emotions give direction to our behavior, without us knowing it, but we can explore them. We can have a slight feeling of anxiety and not know what is causing it (Damasio, 1995).

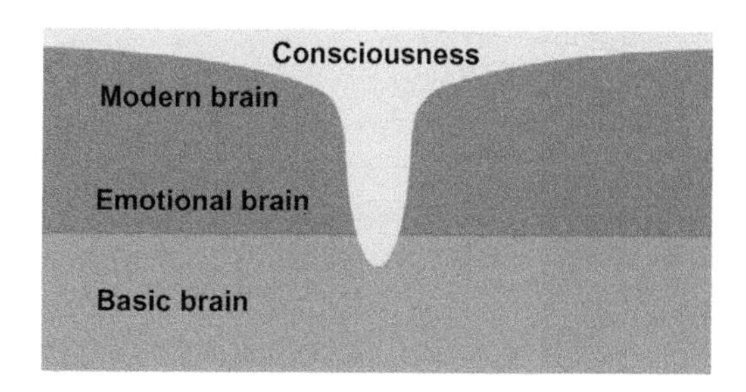

Most of the activities of the modern brain even act on a nonconscious level. When we plan how to drive from one destination to another, we quickly define a route. We can do this consciously when a new route is unfamiliar or nonconsciously when we have driven this route many times before.

Case 2

In case of driving a car, we can become conscious of a familiar route again when we are faced with an unusual traffic jam due to an accident. The modern brain is then needed to define a new route. The emotional brain might be involved on a conscious level when we remember that we recently received a fine because we forgot to adjust our speed to the new limit.

2.8 WHERE IN THE BRAIN?

Readers who are not interested in neuropsychology can skip the following parts without missing crucial arguments for understanding the total message.

We have seen now how three systems of the brain have been developed. Each of the three brain systems is involved in almost every action we perform today (MacLean, 1990). The locus of control can shift from one part of the brain to another, depending on the situation or the task. In very stressful situations, we will easily fall back on our basic brain and act in a primitive way. If everything is under control, we can rely more on the input of our modern brain and actions will become more sophisticated. Most of the activities of the brain happen autonomously and without any interference from our consciousness.

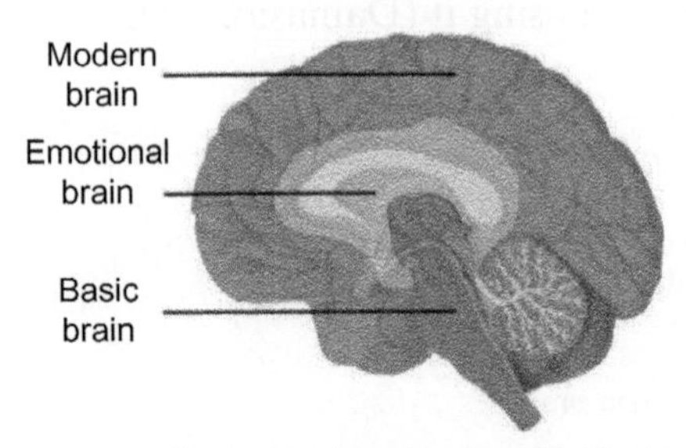

Basic, emotional and modern brain

The systems have a correspondence with certain brain areas, although no area exclusively hosts one system. In fact, many regions in the modern brain that are officially designated as the home area of the emotional brain are also used by the modern brain (Pessoa, 2008; Duncan, 2007).

As far as we can attribute areas to systems, the basic brain is constructed out of the spine (the input and output highway of the body), the brain stem (managing bodily functions like respiration and heart beat), the hypothalamus—including the pituitary gland, also named the hypophysis (regulating many functions like day/night rhythm or sexuality via secretion of hormones), the thalamus (the central control room, nonconscious perception), and the pineal gland, also named the epiphysis (regulating sleep) (Kolb, 2008).

The emotional brain or limbic system is constructed out of areas located in the center of the brain around the thalamus, like the amygdala (responsible for anxiety and attraction), the hippocampus (stores learned information), the gyrus cingularis (safety screening and autobiographic archive), and parts of the frontal cortex (labeling feelings).

The modern brain has its main base in the neocortex, the outside shell of the brain, and is a host of many functions like perception, the reasoning and executive functions, language, and the motor functions. The center of reflecting and reasoning is located just above the eyes. Heavy thinking can even cause a headache in this area.

The pain system is not fully grasped yet. We have pain sensors in the skin, tissue, muscles, and joints. The pain signals are first analyzed in the spine. Special cells can generate reflexes to withdraw our body from dangerous stimuli like fire or sharp objects. Next, these data are transported via separate pathways in the spine. All pain data are collected in the brain stem. Via the thalamus, the signals are sent to the insula in the neocortex. The modern pain center is located here, collecting all the data and relating them to other live events.

2.9 SUMMARY

- We can distinguish three main functions in the brain, each playing a role in safety management.
- The basic brain senses general risks and helps the body to anticipate handling dangers.
    - The basic brain works on a nonconscious level. We cannot directly experience the output of the basic brain, but we will experience hunches that tell us what to do and what to avoid.
    - The basic brain is fast compared to our conscious system. It takes 0.5 second to realize what's really happening and to start a reaction. In the meantime, the basic brain is preparing our body and mind.
    - In case of real stress, the basic brain acts even before the other brain parts can be involved.
    - The male basic brain has developed three basic ways to deal with danger: fleeing, fight, and freeze.
    - Fleeing and fighting are very impulsive: Act first, check later.
    - The female basic brain stimulates care and share behavior: to tackle the danger together.
    - The basic brain is focused on the here and now. It is not a part that plans behavior. The actual stimuli are important. Neglecting the total picture leads to biases.
- The emotional brain supports us in how to relate to the external world.
    - Anxiety is a basic emotion. It is a warning system anticipating pain.
    - The perception of risk is based on a specific stimulus that is associated with anxiety. This association can be innate, but is usually learned.
    - Risk sensitivity warns us of dangerous objects and of unreliable people.
    - The herd instinct uses the knowledge of the group, but can also lead us to collective failure.

- The basic relational patterns developed in tribes are still the foundations of present social behavior.
- We automatically learn a lot from group members, and the small intimate group (in-group) is the best place to gain understanding of safety issues.

- The modern brain is involved in reasoning and planning, the more technical risk assessment.
  - This part of the brain can define a strategy and act according to it.
  - The modern brain can understand possible risks in a plan or project.
  - Risks are calculated and taken whenever needed. Survivors manage equilibrium on the risk spectrum.
  - Males learned to engage in more dangerous situations; females have higher sensitivity to possible dangers.
  - The modern brain is not programmed to avoid risks at all costs. Safety management needs to compensate for this.
  - The modern brain likes to build in a safety margin.
  - We can more easily estimate the risks someone else might encounter compared to the risks we might encounter.
- The brain is fully equipped to handle dangers, although the dangers of today are completely different from the dangers of a few thousand years ago.
- We all have a complete infrastructure for detecting and handling risks, but our innate programs are not adjusted to the present risks. Risk sensitivity needs to be programmed, both in the domestic and work environments. Without this programming, we have no alerts for risky situations.
- The task of parents is to raise their children in such a way that they learn to estimate the real dangers of living in this new society. The same can be said for employers. They also have the task of training their employees to discover and handle the dangers involved in working.
- We have to realize that although we think we are conscious beings, we are in fact acting mostly on a nonconscious level. Safety management becomes effective if we are able to influence our nonconscious processes.

TIPS FOR TRANSFER

Safety Investments Never Stop

The fact that we all have a brain that is programmed with a lot of risk tolerance, especially for circumstances in the 21st century, teaches us that we need a structured and continuous process to keep safety behavior on a high level. Without giving continuous energy to the safety process, it will lose its effectiveness. An organization will never reach a point where it can say, "Finally, we've invested enough to reach complete safety."

From a safety management perspective, one could get irritated or disappointed by the fact that safety behavior has a tendency to decline, regardless of the investments made in it.

Question: What can you do to ensure your constant readiness for new perspectives in safety and safety behavior?

We will discuss many inherited elements of our risk history in this book.

This page intentionally left blank

PART 2

Risk and Safety in a Neuropsychological Perspective

In Part 1, the final issue of creating complete safety is narrowed to managing the risk behavior of employees. Before we go into the question of how to manage this risk behavior, we will first need a thorough understanding of how we deal with risks.

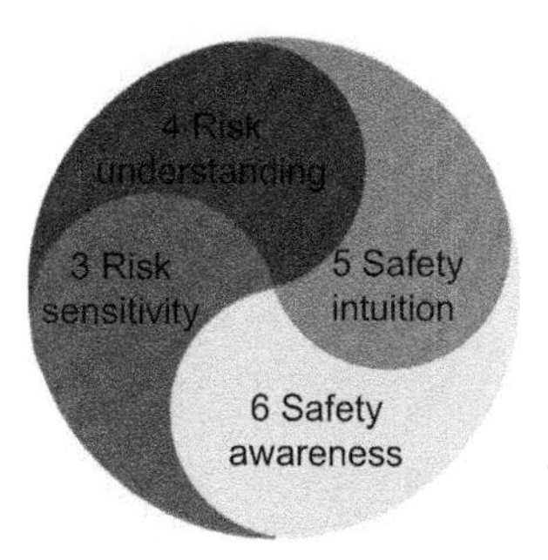

To gain this understanding, we will use perspectives that have recently been developed in psychology. In Chapter 2, we saw that the brain has developed many ways to deal with risks. We will explore these ways in detail in the next four chapters.

We will start at the beginning of many brain processes, the perception of external stimuli. Although the perception process is usually seen as an innate competence, we now know that this assumption is far from reality. What we inherited is a possibility to perceive, via wires that connect our senses to perception areas in our brain. These brain areas are almost blank when we are born. Some exceptions relate to basic functions: survival, reproduction, and cooperation. A newly born child probably easily learns to perceive faces, and by the time we become sexually active, all sexual stimuli are perceived with priority. In all other cases, perception is only possible after a learning process. This process starts in the mother's womb and ends on our deathbed. Once we have learned to perceive something, we can store it, remember it, and use it. The main reason why we remember so little from our early childhood is that we didn't yet have the ability to perceive and store information, so we have nothing to retrieve from our personal archives about those years. Because perception is a result of a learning process, and all learning processes are different, we can conclude that perception is personal. There is no general and identical way in which we all perceive something. Everybody creates his own reality. There are strong resemblances in, for example, estimating speed or size but large deviations in the perception of risky behavior. Transferring this knowledge into the area of risk management, we must realize that almost all risks in modern life have to be learned. The end result of this learning process, the ability to sense a certain risk, is called **risk sensitivity** (see Chapter 3). Well-developed risk sensitivity usually protects us from small incidents.

> Most perception is the result of a learning process. This makes perception personal.

> Everybody perceives risks in his own way.

Once we have stored information about our environment and we know our plans, the brain starts to digest all the information and uses it to scan for possible dangers. This is an autonomous and nonconscious process that literally works 24/7. This process can be enhanced by giving attention to it and especially by adding additional information. The more we understand about the work processes we are dealing with, the better the quality of the scanning. We generate **risk understanding** (see Chapter 4), helping us to take the safe track. Well-developed risk understanding protects us from major disturbances in processes.

From an outsider's perspective, risk sensitivity and risk understanding have a strong resemblance. Without any understanding of brain processes, we would not be in a position to distinguish between them. Neuropsychology teaches us that they are, in fact, two completely different processes, each with its own manual. The end result of these processes is handled in the same way by the same areas of the brain. If these processes do a proper job, we know where risks are and how serious a threat they might pose.

The next step is that we have to anticipate these risks. Again, there are two completely different processes involved with a similar result (Epstein, 1994; Sloman, 1996; Evans, 2003, 2008; Kahneman, 2011). One takes place on a nonconscious level, the other on a conscious one. As the name says, we are hardly aware of the nonconscious one and only experience the conscious. One might ask why it is relevant to have knowledge of a nonconscious process, if nobody experiences it. First of all, it is the main source of our behavior and even if we don't experience this process ourselves, others are directly confronted with our behavior. Second, the nonconscious processes can be far from rational; they suffer from biases that sometimes prevent the safest behavior from being apparent. Fortunately, these nonconscious processes can be influenced, both by us and by others. Due to this sensibility for external stimuli, management can influence their personnel. Although we might think that such an influence is mainly created via the conscious channel, this notion is far from reality. We mostly influence each other via nonconscious processes. Part 3 of this book is dedicated to this last topic.

The nonconscious contributes in a crucial way to our personal safety behavior, without us knowing it.

In Chapters 5 and 6, we discover more about the cooperation between the conscious and nonconscious processes. In literature, this cooperation is referred to as the dual process theory (Chaiken & Trope, 1999), stating that the nonconscious and conscious processes work strongly together.

Chapter 5 is dedicated to the nonconscious process that is called **safety intuition**. This safety intuition takes care of all standard and known safety situations. The automatic processes protect us from at least 99 percent of all the risks we encounter. Only when we are facing new situations and no stored solutions are available does the nonconscious seek help from the conscious system. We call this conscious system **safety awareness**. It cares for our safety in all special circumstances and, in doing this, it feeds the nonconscious system so that it can handle this risk in the future.

The nonconscious system works below the surface of awareness. It is our automatic pilot and uses all the previously learned routines that are put in place whenever needed. Because it uses programs that are lying on the shelf, it consumes hardly any energy and requires little effort. As it is automated, the nonconscious system is faster than the conscious, something that can be very useful in times of acute risk. Some of the processes that run via the nonconscious system are susceptible to conscious control (breathing, chewing, and so on), but they usually work fully automatically. In some cases, conscious control can even interfere with a nonconscious process, for example, learning how to ride a bike. Anyone who has ever tried to explain to someone else how to ride a bike knows that some things have to be carried out in practice but are not as easily explained.

The nonconscious is fast, uses less energy, and takes little effort.

At the moment of birth, there is a strong resemblance between the perception system and the nonconscious behavioral system: Both are almost empty. As newborn babies, we have some innate reflexes such as clinging to our mother or the ability to swim, but these reflexes disappear soon unless they are being used all the time. So both our perception and our behavioral system need to be filled in our lifetime. Nonconscious systems can enrich and adjust themselves automatically to a certain extent due to conditioning and learning. If we think in terms of behavioral improvement, the nonconscious system can handle this without any interference of the conscious system. But if we think in terms of behavioral renewal or innovation, we need the involvement of the conscious system because new patterns have to be acquired. A paradigm shift is always the result of conscious interventions. Another difference between the conscious and the nonconscious systems is that the nonconscious system does not focus so much on detail; rather, its orientation is holistic and it places activities in a wider perspective. Research shows that the nonconscious system is not only faster than the conscious, but even the quality of the problem solving can in some cases be higher (Dijksterhuis, 2007).

Improvement of behavior can be achieved fully in nonconsciousness; innovation requires consciousness.

The conscious system is relatively small but important. It gets involved when the nonconscious system cannot handle the risk. It can gather data stored in our memory; analyze them; discover connections, causes and effects; build theories about unexpected events; and draw conclusions. It can make plans that are usually executed again by the nonconscious. The conscious system is really helpful if we are confronted with new situations. All mental exercises of this system are done in our scarce working memory, which means that the amount of data we can handle consciously at a particular moment is limited. If something else draws our attention, this disturbs the conscious system and the data collection has to start all over again. This makes the conscious system slow. Besides that, it uses a lot of energy. We get tired from thinking.[1] In case of emergencies, the conscious system is probably too slow to take appropriate actions. When we think about ourselves, we usually refer to this conscious part. Consciousness is what makes us human. It takes quite some imagination to realize and also accept that what we do stems mostly from the nonconscious system. Safety awareness is discussed in more detail in Chapter 6.

[1]To get an idea about the energy used, the total brain contains only 2 percent of our body mass but uses 20 percent of all the energy. If we run out of energy (oxygen or sugar), we experience motor problems like a lack of coordination. These problems are not caused in the muscles but in the areas in the brain that steer our motor behavior.

Case 2

In the case of driving a car, we can see how these two systems work together. During our learning phase with a driving instructor, we first start acting on a conscious level because we have no automated behavior yet. We learn the effect of using the gas, the brake, and the steering wheel. While learning these elementary items of driving a car, we don't have additional space in our consciousness to also watch the traffic. The instructor takes us to a spot with hardly any traffic and prevents accidents by using his second brake. Once we gain control over the vehicle, we stop thinking about how much force we need to use while pressing the brake to reduce the speed a little bit. After a few hours of driving, the nonconscious controls these basic skills. The consciousness can now be fully dedicated to learning to drive a car among many others users of the road. The instructor teaches us how to anticipate traffic coming from different directions. We consciously learn the meaning of a car coming from the left and how we have to behave in such a situation. After some time, these behaviors are taken also over by the nonconscious. Now we can use our consciousness to fully focus on the traffic and learn patterns in how to drive and anticipate others, for example, by adjusting the distance between your car and another depending on the speed. Step by step, our nonconscious takes over control, and after some time, we can even forget that we are driving a car. We just make a conscious plan of where to go and the nonconscious brings us there. In the meantime, we listen to the radio or mentally prepare an action at work. There are even recorded situations in which people drive a car while sleepwalking, not remembering anything the next day when they wake up. That is an example of 100 percent nonconscious driving.

In Part 2 we will not only describe the four main processes of our brain, but also the way in which they are connected to anticipating risks and generating safe behavior.

From a rational point of view, one would imagine a process flow that starts with risk sensitivity and risk understanding, which together create input for the next steps: the nonconscious and the conscious system that transform this input to behavior appropriate to the task of coping with risk. Unfortunately, this is not the way the brain is designed. In fact, all the brain activities are interrelated. We cannot perceive an object without checking in our memory if that object has

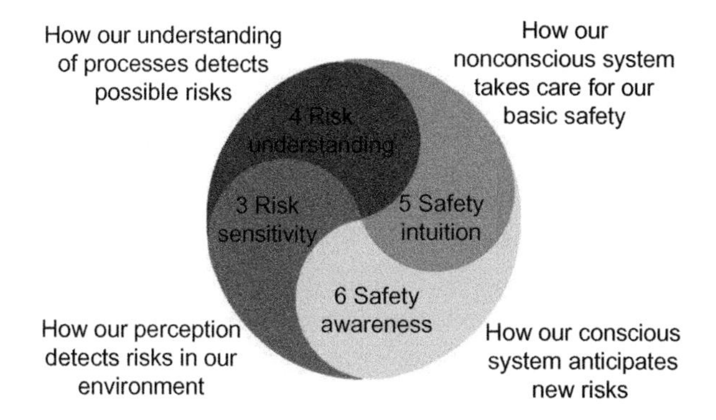

been noticed before, and developing a plan to tackle a risk constantly requires input from our senses. That's the reason why the figure is made circular, suggesting a permanent dynamic interaction between the four elements.

CHAPTER 3

Risk Sensitivity
The Perception of Risks

We start at an interface between our environment and ourselves: our senses. When we have the ambition to go for completely safe behavior and reduce risk from our behavioral repertoire as much as possible, the first question is: How can we enhance the notion that something can or might be risky? The brain uses the danger system to identify these risks. The danger system once started with the pain system giving signals that something has gone wrong and that action should be taken to avoid more serious threats and to recover from previous wounds. The next step was to anticipate possible wounds for which we needed a warning system. Anxiety is the messenger of pain and is therefore a crucial emotion in all safety behavior. We are always anxious for something (for example, an object, a situation, a living being, or a group). Risk sensitivity plays a central role in this warning system and can be defined as an anxiety reaction to a specific external stimulus. The main effect of risk sensitivity is safety intuition and safety awareness, which will be discussed later. Usually, risk sensitivity is active on a nonconscious level.

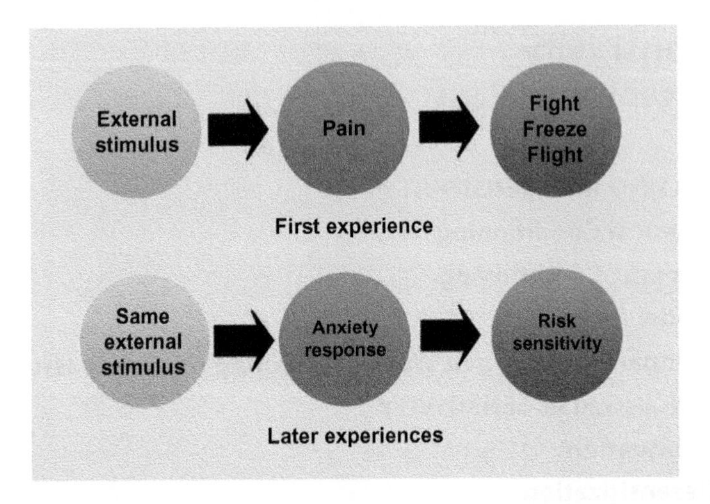

As stated earlier, during evolution, we programmed our natural risk sensitivity (our first nature) in a completely different environment, in the African savannas, which is incomparable with a modern industrial environment. This change in environment has an impact on the quality of risk scanning. For some external stimuli, we can still rely on the process inherited from ancient times, while for most other stimuli, we have to activate the risk sensitivity as an almost artificial process. We have to reprogram this function. The fact that it is an artificial activation of a familiar system is a key element in understanding and enhancing safety policy. The risk sensitivity that we need in modern life belongs to our second nature.

Innate risk sensitivity reactions, like mental alertness, can be seen as responses to familiar threats like:

- Heights and heat
- Insects and small animals like snakes or objects with a remarkable resemblance to them
- Moving objects, especially those moving in one's own direction and generating sounds
- Nonmonotonous sounds (for example, a siren)
- Adverse weather like thunder, lightning, and storms
- New environments and unfamiliar faces

If we examine this list, we can state that some of these stimuli can appear in a modern work environment, but they seldom do. The most interesting one is the newness of an environment, something that is always present when changing jobs. Although it is a general attribute, newness generates a mild sense of anxiety due to the fact that not everything can be overlooked and controlled. As long as a situation is being perceived as new, our overall level of awareness is raised. It is a natural way of being mindful. Newness helps new employees during the first one or two years in a position. After that period, this effect fades away.

Unfortunately, employees that are new on a job are not only more aware of the situation but also display more risky behavior. This is a direct result of the fact that many of the risks in a new job are not on the previous list. There is no innate warning system for these situations; our first nature is insufficient to manage safety in these situations. Working safely requires a second nature on top of our first nature, an artificial risk sensitivity that must be learned for almost all dangerous stimuli.

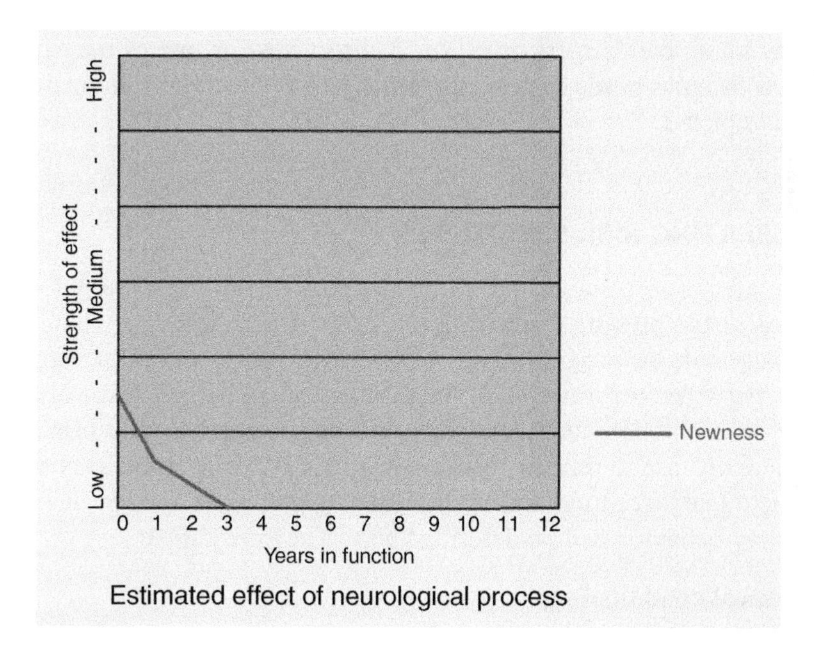

Estimated effect of neurological process

We are all very familiar with this need for educating risk sensitivity when we have young children. We are aware of the fact that they have some innate danger sensitivity, usually connected to certain periods in their development, for example, the fear of being left alone around 18 months or the fear of unfamiliar faces around four years. Everybody knows that we have to teach our children most of the dangers of this world, from the dangers of a hot stove to the traffic on the street, to a sensitivity for avoiding interesting pills at a dance party though the majority of attendants is using them. This teaching is a form of (re)programming. Some elements belong to our innate danger repertoire, like a sensitivity for being more aware of dangerous sounds or objects like moving cars. Most of all, we need to share our concerns and feelings of anxiety, which our children will sense. Step by step, our children learn to recognize and to anticipate the dangers of this modern world.

All employees need to be raised in becoming sensitive to dangers.

A comparable teaching process has to be done when an employee starts to work in a new work environment. Just like raising our children, we have to start from scratch during the induction period and teach new colleagues to become sensitive to all the risks involved in the work, one by one. It is a deadly sin to assume that new colleagues already know the dangers involved. A direct boss or one of the experienced colleagues needs to take the role of safety coach until induction is fully completed.

3.1 CREATING RISK SENSITIVITY

Every safety coach should have knowledge of two things: the potential dangers in the present work situation and the principles by which risk sensitivity can be programmed. A small overview of general learning psychology can introduce us to the different ways we can learn or condition behavior. Humans don't rely on one specific learning principle, and actually use a mix in which a whole variety of principles work together. The three most important learning principles used in developing safety behavior will be described here. They are called:

1. Classical conditioning,
2. Operant conditioning, and
3. Model learning.

3.1.1 Classical Conditioning

Classical conditioning has become famous since Pavlov's dog experiments, and is active in many aspects of our life, for example, in the principle behind advertisements. Pavlov was a Russian biologist researching the secretion of saliva in dogs. He inserted a fistula in the throat of a dog and tapped the saliva while the dog was watching the food. At a certain point, Pavlov discovered that the dog started to salivate before the food was even in the laboratory; the dog responded to the sound of the food server's footsteps as he walked down the corridor bringing the food. The sound of the footsteps had become a trigger to start the saliva reflex. The basic principle of classical conditioning is that an original stimulus (visible food) is connected in time and space with a new stimulus (footsteps), and that the new stimulus can evoke the same reaction as the original stimulus after a while. The central element in classical conditioning is that by perceiving two stimuli together, an association is created in our brain. Due to this association, both stimuli can evoke the reflex that once was only connected to the original stimulus. Once the connection is established, the neutral stimulus can evoke a reflex that originally didn't belong to that stimulus. We call this association a conditioned response (in this case, salivation) to the new stimulus. The connection keeps on working when the original evoking stimulus (food, in the case of the dog) is not active or present anymore. The new stimulus has become a conditioned stimulus.

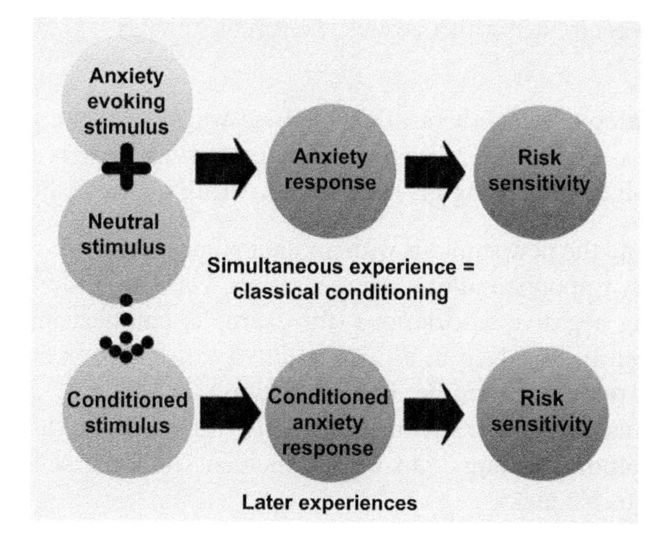

Pavlov continued his experiments and started to ring a bell just before the server would enter the corridor of the lab. After some time, the salivation process started when the dog heard the bell. A conditioned stimulus (the sound of the footsteps) could again be used to condition a second new and neutral stimulus (bell) in such a way that it gains the quality of a conditioned stimulus. Via this principle, the original response could spread over a category of stimuli.

Classical conditioning works not only in the area of physical reflexes but also with emotional reactions like anxiety, pain, or happiness. As mentioned earlier, risk sensitivity is the associative connection of (a slight feeling of) anxiety and/or pain with a stimulus. A safety coach who wants to sensitize a new colleague for a potential risk involved in, for example, handling a tool, needs to create an association between the new tool and a feeling of anxiety. Once this association is established, the colleague will sense his own anxiety as soon as he sees or works with the tool again. We call this learned anxiety a conditioned anxiety. This conditioning process is a crucial step in gaining risk sensitivity. As soon as the new colleague experiences the possible danger of the specific tool (for example, an electrical saw or a sharp knife), all other comparable tools will be experienced as dangerous.

Classical conditioning teaches you to have an emotional reaction as soon as you are confronted with a possible dangerous stimulus.

For a safety coach, there are two basic ways to evoke an anxiety feeling in the mind of a colleague: using an existing anxiety of the colleague involved or sharing his own anxiety. Both are described here:

1. Connecting the new stimulus with another stimulus that already evokes an anxiety response and/or a pain sensation. This can be done via:
   - Adding negative associations (for example, emphasizing the risks that can be involved in a certain activity or environment that are comparable with another known activity).
   - Creating associations with past events (for example, discussing in a "toolbox meeting"[1] an earlier incident that happened with a comparable task).

[1]A "toolbox meeting" is defined as a meeting of all people (employees and contractors) involved in the execution of a task. These people discuss the actual assignment, what actions have to be taken, who is involved for which activity, and what the end result should be. Part of a toolbox meeting is a Safety Risk Analysis, in which all people discuss what the possible risks might be and what can be done to reduce these risks and to solve possible problems in case they appear.

  - Displaying on the spot signs/alerts ("Last year a serious accident happened in this area, while...").
  - Adding anxiety-evoking warning stimuli when dangerous activities are taking place (for example, a special warning tone when a truck is driving backward).
2. Sharing the colleage's anxiety, making use of our ability to feel what others feel (which will be discussed in Chapter 9). This can be done by:
  - Expressing a personal concern for an activity or object, like a warning ("I feel very uncomfortable when I see you doing this, because...").
  - Sharing personal beliefs in the form of statements: "This plant is only as safe as the way people handle it."

Case 1

Classical conditioning helps people develop sensitivity to many dangers on the site, like the steps on the side platform, the heat from the cooler, the squirting oil, and the open pipe.

Case 2

One of the main tasks of driving instructors is to create danger sensitivity. Although most of us have experienced dangerous situations while sitting in a car as a child, we only developed weak danger sensitivity once we start our driving lessons. It is hard to sense the impact of a specific action if we don't fully understand the processes involved in driving. Driving instructors will focus our attention to a specific situation (a car leaving a parking spot, for example), and teach us what might happen, where the possible risks are, and how we can anticipate on each of these risks. The classical element is that instructors share their emotion connected to that specific circumstance, more than rationally explaining the situation.

3.1.2 Operant Conditioning

The second way of learning is called operant conditioning. This principle was originally described by B. F. Skinner, who used hungry pigeons to demonstrate that he could teach them within a period of an hour to peck on a knob, just by rewarding every movement that went in the direction of pecking that knob. This process is called shaping and fully relies on the economic principle behind behavior. The basic rule of operant conditioning is that behavior will get stronger if the perceived consequences of that behavior are positive, and the behavior will weaken or stop if the sum of the perceived consequences is negative. A big difference with classical conditioning is that this way of learning works via the consequences that follow on a behavior, whereas classical conditioning always works by simultaneous stimuli before the behavior takes place. Operant conditioning works by finding out what the effect of a behavior is. It shapes behavior via a feedback loop. If we touch a hot stove, we will feel pain and we will learn that hot stoves can be painful and dangerous if we touch them. Once the feedback loop is established, a shortcut will arise, giving us a sense of danger even without acting.

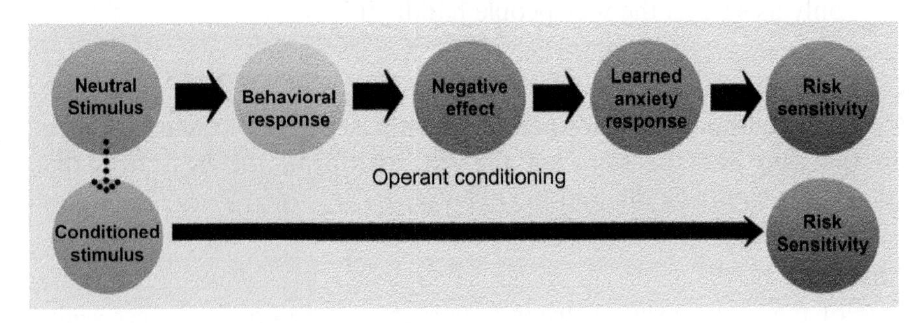

Examples of positive effects (that reinforce the actual behavior) are reaching the aimed goal, earning profits in time (ready to go home) and money (bonus), social approval, and social inclusion. Examples of negative effects (that weaken the behavior) are pain, social exclusion, punishment, and loss of money or goods. Remembrance of previous problems or accidents can have a comparable negative impact on a stimulus. It is important to notice that the words **perceived effects** are used. An effect that is *perceived* as reinforcing for one person is not necessarily perceived like that by another person. So the same set of consequences can stimulate one person to continue or even strengthen a specific behavior, whereas it can weaken or stop the same behavior for another person.

Operant conditioning always requires trial and error.

For new behaviors or for old behaviors in new situations, we mostly do not know the effect in advance. In these cases, operant learning works according to the trial and error principles: learning by finding out what happens after the exposed behavior. This trial and error principle is very effective in many parts of life, but in general, we don't want this to happen in safety management simply because we want to exclude errors. Errors can easily lead to near hits and incidents. There are some exceptions in which the principle of operant conditioning can be used to some extent in safety management:

- Deliberately creating situations in which things go wrong to create sensitivity by giving others the opportunity to find out what can happen. This strategy is only suitable if the possible risks are low and the possible damage to a person and his environment can be repaired easily. (Allowing a child to burn his hand once while touching the stove is a very effective way of teaching him not to touch it again.)

Punishment teaches us what to avoid, but does not reinforce desired alternatives.

- Adding negative consequences, for example, punishing people who have trespassed a safety boundary. Both a boss and his team can play the role of punisher. Trespassing a safety boundary will later be associated with higher risk sensitivity. We always have to take into account that punishment teaches us *not* to perform a certain behavior, but it doesn't teach us what we actually should do. Punishment can even reinforce undesirable behavior, for example, learning that rules can only be trespassed when the boss is not present. For this reason, it is strongly advised to use punishment as little as possible in safety education.
- A much safer way of using operational conditioning is working with simulations, for example, a plant or flight simulator or a dummy (a plastic arm or an actor) in a hospital. Such a simulation can supply feedback for many behavioral options and will be experienced as if the situation is real. The more realistic a simulator is, the stronger the learning effect and transfer of the lesson to real-life situations. A simulator is the only way for a person to experience the outcome of a certain behavior without the damaging effects that would be generated in real life. People using the simulator will incorporate the feedback into their whole perception of the world as if the situation

was real. All pilots that fly you safe around the world are trained regularly in a flight simulator.

Operant conditioning can be effective while using a simulator.

Case 1

Operational conditioning needs consequences, things happening as a result of behavior. Both an operator and mechanics learn after an incident (for example, what to do next time). This learning is strong, although it can only become effective as a consequence of discovering the cause of the incident.

Case 2

The feedback one receives from a driving instructor is a strong reinforcer. A car simulator, although hardly used, provides perfect feedback. The sound of the distance sensor while driving backward gives perfect feedback about the actual position of the car.

3.1.3 Model Learning

The third learning principle is called model learning, learning from each other. This principle is active at moments we don't act ourselves, but when we see others act. The other person becomes a model and, due to a mirror system in our brain (we will discuss this principle in depth in Chapter 10), we can integrate perceived behavior into our system as if it were our own behavior. Model learning is active on a nonconscious level and can work for all actions that we are able to understand and/or that are already part of our behavioral repertoire. So in model learning, we mirror the action of somebody else as if it were our own action and learn from this action as if we were performing it ourselves. We learn by observing, because we feel what others feel and we think what others think. If, for example, we see an accident happening, we can feel the pain a person has and by feeling this, we create a link between a stimulus and danger involved. Model learning is self-reinforcing because it is intrinsically rewarding to belong to the group. People who copy each other's behavior are experienced as more sympathetic and attractive as group members.

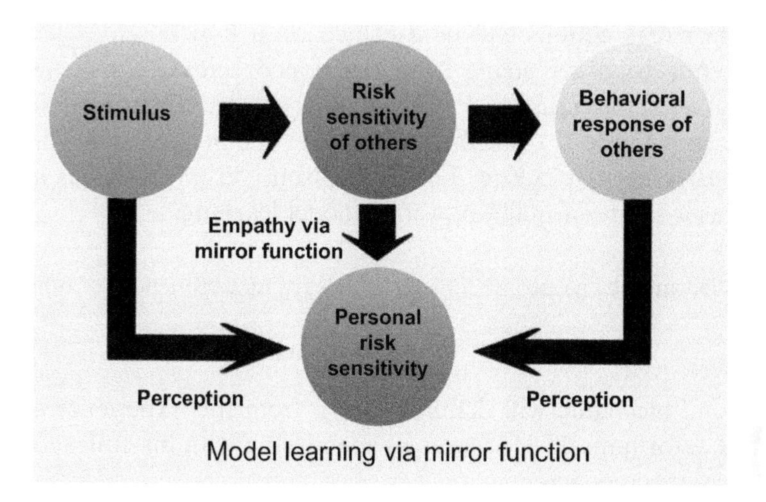

We can recognize examples of model learning in safety issues:

1. During the induction period in a certain job, senior employees can play the role of safety coach just by acting safely and by vocalizing their thoughts (thinking aloud) while doing the job. Assimilating modeled safety issues will be easier if the intention of the expressed behavior is better understood.
2. When people share narrative experiences (what they have seen, heard, or experienced while working), these experiences and connected emotions are planted as an association in the brain of someone else. The emotional impact of these narrative experiences is contagious; it jumps from person to person.
3. People will model people in films or videos who could be colleagues and who display a certain safe procedure if the models are attractive.

The general rule is that a more attractive model leads to stronger model learning. The higher a model's perceived status, the stronger the tendency for others to learn from them. Members of management and informal leaders within a team have more modeling power compared to other team members. A second rule is that people follow models for their actions in combination with their hidden intentions. Fake behavior, in which someone is acting, will not be copied. A third rule is that people are less sensitive to expressed words from models, as compared to perceived actions. Words are usually perceived via the conscious system, whereas actions can be executed on a nonconscious level. As most of our behavior stems from the nonconscious, a message that arrives in the conscious level will only start to be effective as soon as we have transferred it from the conscious to the nonconscious domain. This transfer needs practice. For this reason, verbal messages are not reliable as a self-reinforcing power of model learning.

Model learning helps people to adopt the learning experiences of others.

Case 1

The second mechanic will definitely learn from this experience as if he were the actor himself. He learns by identifying with his colleague.

Case 2

A good instructor will refer to the behavior of other drivers on the same road who can serve as role models. Once a student is developing a frame of reference of good driving, she will also learn from others while being a passenger.

3.1.4 Comparing the Three Ways of Learning for Safety Issues

Although the second way of learning—operant conditioning—is a very powerful one, the fact that it always needs a direct experience with stimuli makes it less suitable for safety issues. Operant learning always involves trial and error, which can easily lead to uncontrolled and risky behavior. Error should be excluded as much as possible, so operant learning should be avoided as much as possible.

Classical conditioning always involves an active addition of a conditioned stimulus to the original neutral stimulus so that this evokes a feeling of anxiety. It can create strong effects, but it requires permanent monitoring of behavior, at least during the learning period. A personal safety coach can provide this input.

Model learning can happen without direct behavioral consequences but just by observing others. Besides that, model learning is almost self-reinforcing because sharing similar behavior in a group makes a person more sympathetic in the eyes of others. Remember that belonging to a group is one of the three basic drives in our life. So mirroring safety behavior of team members is intrinsically rewarding.

> Model learning should lead in safety education and can embed classical conditioning. This defines safety learning as a social event.

In practice, acquiring safety behavior will mostly rely on model learning. The consequence is that the existing behavioral patterns in a group or team have a huge impact on the behavior of new employees. These patterns are "contagious." Good patterns strengthen themselves; bad patterns are like weeds—hard to get rid of because they pop up all the time.

All together, personal coaching and social control of the team will be key drivers to strengthen sensitivity for risks. If we can influence the social control, we can influence risk sensitivity.

3.2 REDUCING RISK SENSITIVITY

For both innate and second-nature risk sensitivity, it is possible to manipulate the strength of the reaction evoked by external stimuli. So far, we have only discussed strengthening this response, but reducing it is also possible. In fact, weakening is a natural process directly related to the amount of exposure to a stimulus. The more we experience a stimulus, the less sensitive we become to it. We call this habituation. There are also cases in which we would really like to become less sensitive, for example, if we have experienced a trauma. In these cases, we can use a technique called desensitization. Both are described here.

3.2.1 Habituation

Habituation is a gradual decrease of strength of risk sensitivity due to:

- Repeated exposure to stimuli, which creates a sense of familiarity. The more familiar we become with stimuli, the more we start to like them, the more previous hostile or anxious emotions will be compensated and fade away.
- Increased expectancy around stimuli, which creates a sense of control. The brain knows what is going to happen and does not need to anticipate so strongly anymore. The need for awareness gradually decreases. In this way, we can become complacent to risks that were previously being perceived as hazardous.

The result of habituation is that the impact of the anxiety response weakens. All processes that usually follow, like safety alertness, will also be activated with reduced strength. Although the dangerous stimulus stays the same, we stay more relaxed. Habituation occurs while perceiving all sorts of stimuli. The more we are exposed to stimuli, the stronger the process of habituation will be. Because habituation depends on the amount of stimuli, it is also related to the spectrum of situations in which stimuli occur. Stimuli that are both present at work and in private life have a higher tendency of becoming habituated because they have more exposure.

All sensory processes are liable to habituation. We constantly lose the freshness of newly learned experiences.

Habituation is one of the main reasons why the risk sensitivity at home generally is so low. A lot of accidents happen in the domestic environment because we are too complacent. The fact that 50 percent of serious accidents occur in the domestic environment is a direct result of habituation. An example is boiling water that we use daily while cooking food or making tea. Boiling water can be very dangerous and can even kill children. Due to daily habituation, boiling water gradually loses its ability to evoke risk sensitivity. At work, a fair amount of accidents occur when boiling water is involved. We can recognize the same with tools that are used both at home and at work. We all know examples of a sawyer in a sawmill who has lost some fingers. This usually happens to those who have many years of experience and in the meantime have lost their risk sensitivity. One small distraction can be enough to saw not only the wood but also the fingers. Usually such an accident precludes the end of that profession, so one would expect that every sawyer is very alert to this potential mistake. The fact that it still happens shows the strength of habituation.

Habituation constantly challenges alertness and eventually tends to win

Case 1

Habituation could be the issue because the mechanics were so familiar with the case that they didn't hesitate for a moment before proceeding with the job, despite not referring to the papers that had been left in the car.

Case 2

Most accidents in the lifetime of a car driver happen during the first five years after getting a driver's license, usually due to a lack of control over the car or the inability to "read" a traffic situation. Most accidents in the following period of five to ten years are caused while doing dangerous maneuvers, like takeovers. Although these activities usually create high-risk sensitivity during the first years, some drivers feel more and more comfortable with handling takeover maneuvers. They have done it so often, and although they agree that it looks dangerous, they are actually convinced that they can manage the situation. This misplaced sense of control reduces the impact of the anxiety system on other brain functions. Habituation has reduced the risk sensitivity.

We can challenge habituation by changing stimuli, for example, changing work, changing colleagues, or changing the environment. Even if a few stimuli are changed, the environment will be seen as new again.

3.2.2 Desensitization

At points in life, some people are too sensitive to danger connected to certain stimuli, and they develop a counterproductive reaction that we call a phobic response. Such responses can be so strong that they can block normal occupational behavior. The person in question can have a strong urge to avoid a specific stimulus and can become so anxious that he cannot handle it in a proper way anymore. We can encounter hypersensitivity after serious accidents with a traumatic outcome. In those cases, it can be necessary to reduce the anxiety-evoking strength of a stimulus through coaching or even therapy. This process is called desensitization. Forced habituation by exposing a person to hypersensitive stimuli is the most effective way to reduce sensitivity.

Hypersensitivity can be reduced by forced exposure to a stimulus.

Case 1

The mechanics may become hypersensitive to this kind of work due to the shock he experienced afterward. The operator could also start to feel extremely guilty because he didn't take all possible precautions (for example, he left the papers in the car and didn't take the two-minute walk back to the car to retrieve them). In such a case, a few sessions of coaching might be needed.

Case 2

Suppose a student witnessed a collision of a car and a truck just in front of him a few years before he started his own driving lessons. He saw the crush, the car driver being catapulted, a lot of blood on the street, an ambulance coming, and so on. Now that he is driving a car himself, he becomes extremely anxious each time another car driver passes. The driving instructor needs to support this student to desensitize the experience before he can become a good driver and pass the driving exam. A very precise description of the memories can help the student to become less sensitive.

3.3 THE COMBINED EFFECT OF NEWNESS AND SENSITIVITY

The combined effect of the learning/conditioning, together with habituation, can be as displayed in this graph: starting low because much sensitivity still has to be learned. During the first five years, there is a steep learning curve. After this period, habituation wins territory and employees becomes less sensitive to stimuli again, although they know the importance of them.

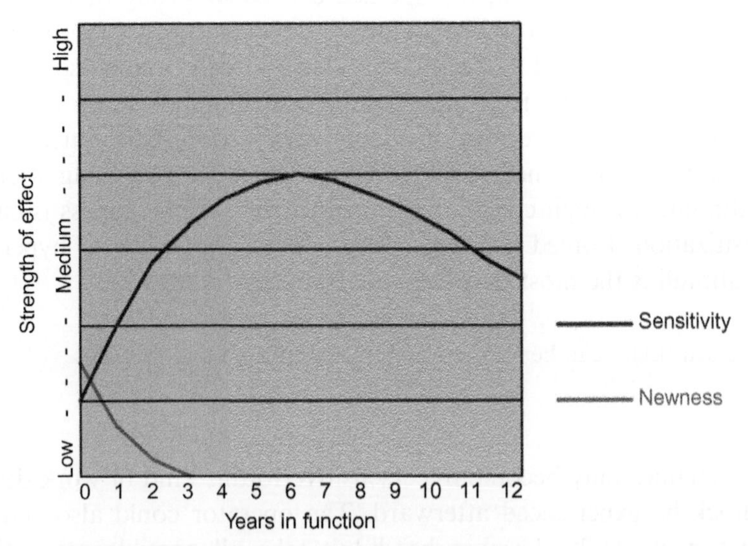

Estimated effect of neurological process

Case 1

The mechanics may have been doing this for many years already and habituation may have started to reduce their sensitivity.

Case 2

A main source of driving accidents during the first years of driving is a lack of risk sensitivity, not recognizing where the dangers are (the other main source is a lack of driving skills).

3.4 WHERE IN THE BRAIN?

All input from our senses (seeing, hearing, touching) first stops in the thalamus (our central information room).[2] A main task of the thalamus is scanning all input for possible danger. The external input passes through a safety check before it is processed in other parts of the modern brain. The amygdala, which produces the basic emotions (Murray, 2007), labels some of the stimuli as potentially dangerous (LeDoux, 2002; Pessoa, 2008). This is done with the help of the modern brain, which gives meaning to certain stimuli, the programming as possibly dangerous (Pessoa, 2010). Next comes the hippocampus, whose main task is to organize the process of storing learned experiences. The brain has archives all over the cortex, but the labels for risk sensitivity are stored right next to the hippocampus in an area that is called the perirhinal cortex. As you can see on the illustration, these all are positioned in the center of our brain. The perirhinal cortex is the storeroom of all sensory information related to objects and the potential consequences that might be connected to these objects (Barbeaua, 2005; Devlin, 2007; Chaumon, 2009). Later, we will see that this information can be sent for further processing via the cingulated gyrus to the prefrontal cortex.

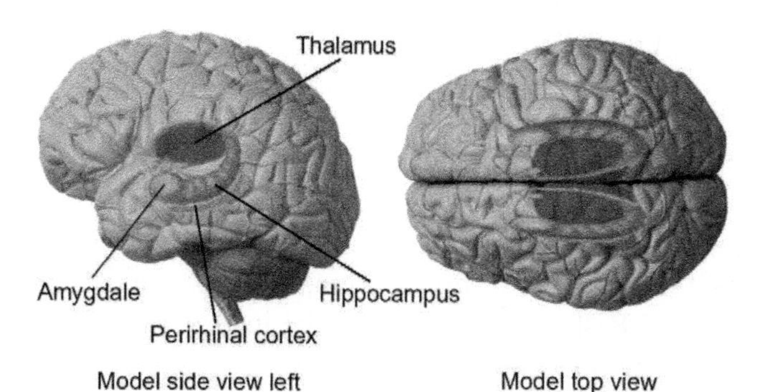

Danger system

[2]The sense of smell is the only exception. The smell input from the nose first enters the emotional brain before it is sent to the thalamus for integration with the input from the other senses. That's the reason why smell has more capacity to evoke emotions like anxiety, aggression, attraction, and so on. This topic will be addressed later.

3.5 SUMMARY

An important element of our survival kit is risk sensitivity. We get it for free, but unfortunately, this sensitivity is programmed for stimuli that are only partly present in our work. The most important one we can use daily is extra sensitivity for all new things and situations. Other danger sensitivities have to be learned. We have three main ways of learning. The first is the classical conditioning that connects a neutral stimulus with a fear-evoking stimulus. The second is operant conditioning, learning by experiencing the effect of handling a stimulus. Model learning is the third one, learning by seeing others act. The trial and error aspect of operant learning makes it less suitable for safety learning whereas model learning is most powerful. That's the reason why social context is very important in generating more safe behavior at work. The biggest enemy of safety is habituation, gradually getting used to (dangerous) stimuli just by being confronted with them. Employees who start in a new job have a high but unfocused risk sensitivity. Although focus grows, the general level of sensitivity decreases.

TIPS FOR TRANSFER

Tip 1: What Needs to Be Programmed in a Safety Conditioning?

Take one process (task) or one physical situation (for example, a part of a plant). Define what stimuli need to be conditioned so that an employee has all the needed risk sensitivity to do a safe job.

Later, in Chapter 7, we will see how a safety buddy can use this list in coaching a new employee to become sensitive to possible risks.

Tip 2: Safety at Work, Safety at Home

As mentioned in the general introduction, safety management involves a paradigm shift. Risk sensitivity has to be incorporated into the deepest parts of our brain, into our perception.

For some areas of life (for example, a dress code: tie at work, jeans in the weekend, and naked in the sauna), it is possible to use different sets of standards in different situations.

Other than our clothes, we cannot change our risk sensitivity when we leave work. It is a conditioned process. Although some dangers are completely different at work and at home (machinery, chemical substances), other dangers are exactly the same (boiling water, steps, cars). The reaction to potential risks is conditioned, and this conditioning is mostly independent of the situation. For the brain, the difference in location (home versus work) only has a small impact on the conditioned perception of risks.

If a company wants to reach a completely safe environment, gaps between safety behavior at home and safety behavior at work have to be reduced. Standards for safety behavior at home have to resemble the level needed at work. Considering the fact that so much first aid is delivered to people who experience domestic accidents, we can conclude that risk sensitivity at home in general is far too low and needs to be upgraded.

Question: What can you as an employer do to increase the level of risk sensitivity and safety behavior at home?

Question: Would it be possible to organize "safety at home" workshops for the whole family, to create safety awareness at home that equals the needed safety awareness at work?

Question: Would it be an option to distribute a safety kit for enhancing home safety (smoke alarm, fire extinguisher, safety glasses, hear protection, and so on)?

Tip 3: The Impact of Safety Messages

Of course, we want to stress the fact that the employer does everything possible to improve the safety of the plant. This is very relevant, especially for external communication. But if we consider the example of driving a car: Driving can never be safer than the behavior of the driver, no matter how safe the car might be. The same is true for a company. A plant can never be safer than the way it is operated.

Being very positive about the actual state of safety management creates an unwanted comfort zone that reduces risk sensitivity overall (classical conditioning). Although external communication can be positive, internal communication should continuously stress the fact that safety relies on the strength of the weakest chain.

Question: Any suggestions on how to handle this topic?

Tip 4: Job Rotation

Long-lasting habituation kills risk sensitivity. Only a few people know the art of looking at work in a fresh way every day. Job rotation is one way of avoiding an impassive attitude toward work. A new position refreshes employees' experience of the environment; it is a good way to postpone the wood sawyer effect.

Question: Can you make an inventory of all employees in a certain job (for example, all mechanics) and estimate the degree of impassiveness they experience toward their work?

Question: How realistic is job rotation in your company, and what would be a reasonable time to shift from one plant to another?

CHAPTER 4

Risk Understanding
Knowing Risks

For the second topic, risk understanding, we leave the area of perception and explore the way we process all the information that is sent to the brain. We experience the results of risk understanding, for example, when we are having dinner at home and suddenly, out of the blue, we realize that we have forgotten something important. This can be an omission in the preparation of a project, a forgotten promise we made to a colleague, or a new insight about what has happened. Usually such an insight comes with a sense of unrest to restore what has been neglected. In most cases, this sudden insight was not triggered by any actual stimulus. This indicates that risk understanding works as a permanent side program below the surface, and we have good reason to assume that it is active all the time, day and night. Some neurobiologists claim that this brain activity is the main reason why we need sleep. While sleeping, we don't have any input and can digest everything we experienced during the day. The brain area that is responsible for this scanning and preparation activity never sleeps.

Risk understanding supports us permanently, day and night.

Risk understanding can be defined as a permanent, autonomous, and nonconscious scanning of activities for possible dangers. The scanning can be done in actual situations (for example, in the workplace), on actions (the past or present activities), and on plans (scheduled work). The process mostly takes place on a nonconscious level, but conscious interventions, like making a project plan or reviewing a checklist, will enhance this process. Risk understanding is directly related to the amount we grasp the situation and the impact of actions that are going to take place. We can only anticipate what we know and understand.

The more you comprehend the process you are dealing with, the easier it is to develop risk understanding.

The difference between risk sensitivity and risk understanding is that risk sensitivity is based on responses to external stimuli, while risk understanding is the result of analyzing processes or activities. Risk sensitivity is active in the here and now, and is always based on previous experiences, whereas risk understanding can be related to the past (I forgot something), the present (is this alright?), and the future (if we don't adjust this plan, something might go wrong). Compared to risk sensitivity that is based in the emotional brain, risk understanding is a more advanced danger system that uses the intellectual power of the modern brain. It strongly involves elements of reasoning.

We experience the results of risk understanding at moments when we receive an alarm signal, usually when we realize that something is going wrong or might go wrong. Typically, such a signal is the sudden feeling of shock or a shiver going through our body, waking us up and making us ready for action. The body physically prepares for possible action. The blood pressure raises and the amount of sugar in the blood increases. When we hear our alarm bell ringing, we can block other planned activities and generate enormous concentration on a specific task.

4.1 ENHANCING RISK UNDERSTANDING

As mentioned, risk understanding is an autonomous and mostly non-conscious activity of the brain. Four aspects influence the quality of this process:

1. Time and timing
2. The amount of data to scan
3. The amount of conscious attention
4. The understanding of the data

4.1.1 Time and Timing

The scanning process starts as soon as one intends to do something, for example, when accepting an assignment. This can be hours or even days before the actual task has to be executed. A toolbox meeting can be seen as one of the starting points for the brain to prepare for the next task. The task description has the function of an assignment that is given to the brain and executed on a nonconscious level. This automated assignment needs an amount of time that is directly related to the amount of data involved. Our brain needs more time to digest big plans. The process also needs some unoccupied brain cells, so it works best at free moments, for example, during breaks at work or during a good night of sleep. Neuropsychologists claim that the basic function of sleep is to offer the brain the opportunity to store the learning experiences of the previous day and to prepare for tasks of the coming day. When you are studying a new language and rehearse some new words an hour before you go to bed, the amount of words you can still recall after one hour is lower than the amount you can remember the next morning. During the night, a lot of connections are established in the brain that help us to store and retrieve the most important information. If we have complicated questions to answer, we'd do best to sleep on it, making decisions the next morning (Dijksterhuis, 2007). During the night, the brain defines the best answers, and we will know what to do once we wake up. So if we have an option to do the toolbox meeting before a break or even as a last action the day before, this will certainly help the brain to be maximally prepared. Doing nothing allows us to prepare for doing something complicated.

A toolbox meeting supports the brain to anticipate what might happen.

We might have a concern that when we plan a break or even a night between the toolbox meeting and the actual execution of a task, the actors might have problems remembering the (safety) details of the task. For this reason, it is good to refresh all safety-related elements in a Last Minute Risk Assessment (LMRA)[1]. As the name implies, the LMRA is done just before the execution of a task, preferably on the spot. Recalling the safety elements once more ensures they are handled with priority while doing the task.

A Last Minute Risk Assessment (LMRA) is a final check to see whether all possible risks are acknowledged and precautions have been taken.

Later, we will discuss the LMRA's contribution to enhancing safety behavior in more detail.

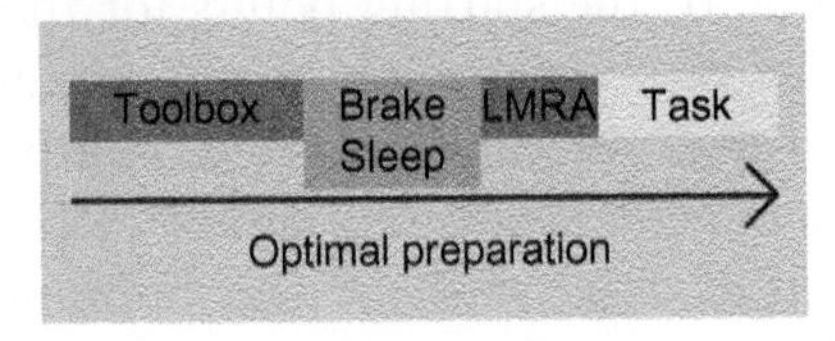

The ideal way to prepare for a complex task in which many people are involved is to first organize a toolbox meeting in which all the ins and outs of the task are discussed (technical preparation), and then take a break or an evening off. As a last preparation, all people involved conduct a LMRA (safety preparation) just before the actual task starts.

Case 1

Any idea what caused the problem in the case study? A problem usually has more causes that coincide at a certain moment. In this case, there were three independent problems:

1. The operator was new on this plant and could not contribute to a toolbox meeting, which was not held because there was no input other than the knowledge of the two mechanics.

[1]A Last Minute Risk Assessment (LMRA) is a meeting on the spot of the action with all the people who are directly involved in the execution of a task. It is a last check in which all the possible risks are assessed once more in combination with the possible activities that might be suitable in case an unforeseen risk might occur. Usually the LMRA is supported by a checklist of, for example, 10 questions to help the people check whether they understand the task, know the safety procedures, and so on. More practical information about the LMRA is presented at the end of this chapter.

2. The mechanics started working on the wrong cooler (so the cooler was still functioning and under a pressure of 5 bar), because the operator forgot his papers in the car and thought he had heard PSG-45D, which hardly differs from what the shift leader said: PSG-45E.
3. The pipe leading to the valve was blocked with dirt. This pollution could gather and stay there because the pipe and the valve are located on the edge of the cooler and it is more troublesome to access that point with cleaning materials.

In this case, they held no toolbox meeting, although there was a perfect opportunity at the moment that the shift leader inspected the working papers of the contractors. He could have organized a short meeting in which he was the chairman and in which the mechanics, a representative of the control room, and the operator were involved. They too easily assumed that everybody knew the task by heart, and they conducted no check on whether the people involved had the same ideas on the task involved. Part of the cause of the problem in this case was that there was hardly any communication between the operator in the control room and the mechanics. If this communication had occurred, the preparation and the actual task might have fit better. Besides that, the LMRA was done too quickly, without the people involved really checking each other's answers. In a thorough LMRA, the involved parties would have taken this possible problem into account. The mechanics were also too accepting of the fact that they hadn't seen the papers and had even failed to register that fact on the LMRA paper.

Case 2

If the driving instructor explains at the end of each lesson what he is going to do during the next lesson and what kind of problems will be encountered in that lesson, students can prepare to a certain extent (for example, by reading a specific part of the traffic regulations).

4.1.2 The Amount of Data Involved

The amount of data varies with the complexity of the assignment. The more specific the plans that are prepared, the more "material" there is that can be scanned, and easier for the brain to gain insights, anticipate risk, and develop risk understanding. Sharing in-depth plans with all participants helps to fill the database of all those involved. It makes work safer. When projects become more complex, the amount of time needed to scan all the material is linear related to the amount of data involved in that project. People need time to digest big projects. In complex projects, it is even advised to split the toolbox meetings, so that the participants can digest the material at their own pace.

4.1.3 Attention

Risk understanding is mostly active on a nonconscious level and guides people unnoticed safely through a day. An occupied mind will have more topics to scan. These topics are handled in a parallel manner. The conscious part of us can contribute to this nonconscious process by giving it attention from time to time. This attention will place the task higher on the priority list. If you are worried about the outcome of a project, you will automatically give more attention to it.

Compare this process with the effect of an agenda for a meeting. If you receive an agenda in advance, you will give it a cursory glance. The next time you see it will be when preparing for the meeting and packing your stuff, then again at the beginning of the meeting, and finally when the topic is discussed. At each of these moments, a little attention is given to the topic. The brain can prepare itself, and the meeting will be more effective. The same process can be used in generating risk understanding. A few brief moments of attention spread over time increase the strength of the process. This topic will be readdressed in Chapter 7 on safety awareness.

4.1.4 The Quality of the Understanding of Data

This topic is one of the most complex elements in the theory of brain-based safety. We can only understand it if we have deeper knowledge about our perception process. First, we will make a side tour in which we study people who suffer from a perception disease called cataract, and next we will translate this knowledge to risk understanding.

Eye lenses that slowly become clouded characterize the disease cataract. People with a cataract can only see very vague shades; they are no longer able to see objects, forms, or depth. Usually, a cataract develops in people between the ages of 50 and 70 years old. There is, however, one exception: innate cataract. People with innate cataracts are born with troubled lenses. They never have been able to see in a normal way and to recognize anything. They can never see more than the difference between light and dark. What is special about this group is that they haven't been in the position to learn how to use their eyes. In the middle of the 1990s, eye doctors developed a new laser technique by which they could cure some of these patients within an hour. So, after having been almost blind for 40 years, these patients were suddenly able to see. After 40 years, the cataract vanished from one moment to another. The main question here is: What does such a patient see as soon as the bandage is removed from the eyes after the laser surgery? Will he say, "Doctor, I'm delighted to finally see you"?

The reality is that these people see almost nothing in the beginning. What once was vague is now suddenly sharp. Unfortunately, the eyes do not even know what it is to focus on an object. Once they learn to focus, they see sharp colors and lines but they still have no idea what they see. After a few days, they find out that some colors in a specific pattern belong together. Two black dots, each with a colored circle around them and some dark lines on the top and the bottom, always seem to move together. Later, they discover that these are the eyes of a person. Step by step, they learn that the pattern is more complicated and that other things like a mouth, nose, and ear also belong to the face. They learn to differentiate between foreground and background. Once a face is recognized and stored in our database, it is a difficult job to find the differences between two faces, especially when these faces change all the time, not only in expression but also in position. For the eyes, this is much more complicated than the children puzzle of "finding the 10 differences." It takes about a half year to recognize friends at a party. Even after years, when these people do a face recognition test, they score badly compared to people with normal sight. Now let's translate the cataract example to the cases in this book.

When growing up, children learn how to perceive.

Case 1

A pump regulates the pressure of the oil in the cooler. Experienced operators can "read" the sound that is produced by the pump. Based on what they hear, they can estimate whether there is fluid or gas in the pipe, how high the pressure is, and the quality of the parts of the pump (for example, when the bearings need to be replaced). A new employee on the job probably only hears that the pump is working and cannot interpret that sound any further to gain extra information from it. A new operator will slowly learn to distinguish different sounds that the pump produces under different circumstances. This learning process can be enhanced if a senior colleague teaches the new colleague step by step to learn to understand the processes in the system and how he can relate the sounds generated by the pump to these different processes.

Case 2

A young driver still has no frame of reference by which she can understand the subtle signals that other road users send. For example, when a car in front slows down, it might mean that there is an obstacle on the road, that another car which has priority wants to enter the road, or that the driver doubts whether he has arrived at the place of destination. Usually, we can find out what really is the case if we combine the car's slowdown with other cues. The better we understand what's happening, the better we can anticipate possible situations and avoid accidents. A driving instructor can help students gain this frame of reference.

What can we learn from this cataract story and the two examples? First, we must realize that what we now see, hear, and smell feels so natural, that we cannot imagine that others may not sense the same when they are confronted with the same input. For us, there is no difference between the way we sense the world and what we call reality.

Second, it is hard to imagine that somebody else perceives the same world in a different way and also calls that his reality. Even when our understanding of the world grows (for example, the meaning of the sound of the pump) and we gain a richer perception, we tend to forget how we perceived the world earlier. So if you image your own team discussing a problem and see your colleagues, each with a different background and experience, you must realize that they all have their own view on a problem and all construct their reality about what's going on. This can make a team discussion troublesome because you don't understand why they don't understand. On the other side, it can also be very interesting to share the different views on reality and to learn from each other's perception.

> We can only perceive what we know.

If we translate these principles to work, we have to realize that everybody who starts in a new job has not only to learn to understand but also to perceive what is there. In the beginning, when everything is new, hardly anything has a meaning. The problem with things that don't have a meaning yet is they can't be stored into our memory or integrated in our thinking process. If a coach explains a part of the process and a newcomer has not yet developed the frame of reference to understand what is explained, he will hear the words but not the meaning and he will forget the whole story. That's the reason why coaches can become desperate sometimes, because they thought that they had explained it well and the newcomer still isn't able to remember it. Step by step, the frame of reference of a newcomer will grow, as will her possibilities to learn. During the induction period, we cannot assume that new employees understand the processes and the risks involved. We need to instruct them because they are starting from scratch. Once they start to understand the processes, they will be able to estimate the risks involved at work.

Newcomers develop a frame of reference, step by step, with which they can understand the work.

So the quality of understanding the process is crucial in risk understanding. Our ignorance for blind spots is one of the most disturbing factors we encounter in this process. This ignorance is two-folded: We don't recognize our own ignorance, and we also have problems recognizing the ignorance of others.

Our own ignorance is expressed in that we don't know what we don't know. Sometimes we can have a sense that we are missing pieces of the puzzle, but usually we don't. There are always holes in our perception, and these holes are strangers to us. The brain seldom has all the information, although we really would like to have it. Like in the old days in the African savanna, the more we understand the situation, the higher our chances for survival. Our basic way to deal with these holes is to fill them in ourselves with assumptions. We simply create an explanation so that our story seems complete again (Dobelli, 2011). But we have to realize that when two people miss the same piece of information, each of them will fill it with her own assumption. It is one of the main sources of misunderstandings.

We don't know what we don't know, and we integrate the elements that we do know into the most obvious explanation.

Case 1

When a little bit of fluid and gas comes out of the pipe a few seconds after it is opened, the mechanic makes an assumption about what's happening. In this case, he thinks that this is just a little residue of the previous process and that there is still minor pressure in the system. Without the second assumption, he would have closed the pipe immediately and would have checked in a different way whether the system was really disconnected.

Case 2

If a driving student is assuming that the car slowing down in front is a sign that it has reached its destination, she neglects the possibility that something might have happened in front of that car that needs to be avoided.

A second form of ignorance is that others don't know what we don't know. Good communication between people is only possible if both the sender and the receiver possess the same pieces of the puzzle. Communication is distorted or not possible if a sender talks about a piece of the puzzle of which the receiver has no knowledge of. In this case, the receiver will either not understand the sender or will assume that the sender is talking about something else without noticing the miscommunication. In a normal meeting at work, this can and probably will happen many times. So a colleague can have an idea that a certain task has been explained while the person involved still has no understanding.

> We can never assume that others know what we know.

Case 1

The shift leader assumes that the day shift operator understands the process, knows which coolers are prepared for maintenance, and has followed the communication in the control room while shutting down the two coolers. Because the operator is new in his job, he does not yet understand the codes that refer to the specific coolers. He just missed a crucial piece of the puzzle.

Case 2

The driving instructor might assume that the student is already so confident in handling the car that she can spend her full attention to follow the other traffic on the road and to listen to instructions. In fact, the student might still be so occupied with the driving itself that she hardly has any mental space left to digest the information and advice from the instructor.

4.1.5 Getting Out of the Vicious Circle

The critical mind will now oppose: How can we learn to perceive? This learning process seems to be caught in a vicious circle. If we do not know that something exists, we cannot perceive it yet and if we cannot perceive it, we can never learn to know it. The solution is to study how children learn to perceive. Actually, they make many very small steps in a circular process during which their concepts of the world grow in an organic way.

> Three steps in perception:
>
> 1. Learn that it exists
> 2. Learn to distinguish it
> 3. Learn to recognize it

Three elementary steps are needed in learning to perceive:

- Learning that something exists (a face),
- Learning to distinguish it from other things (a face of a doll), and
- Learning to recognize it (the face of a particular person).

Once we can recognize it, we can start anticipating it. Employees new to a plant and employees from contractors will lack a lot of information and will have a higher tendency to create dangerous situations because they don't understand and/or do not know that they do not understand. Due to that, they will miss meaningful parts of communication between colleagues. Although colleagues think that information has been given, they forget that the receiver is not yet ready to receive it.

Case 1

When the mechanic arrives in the area of the cooler for the first time, he will have no idea what's really going on. Later, he will experience that the radiated heat from the pipes around the cooler is an indication of the activity of the cooler and an indirect indicator of pressure in the system. As experience increases, he might even be in the position to estimate how long a cooler is disconnected just by touching a few pipes and estimating the heat. The temperature of the pipes then begins to reveal information about the system.

Case 2

Once a young driver has received his driver's license, he will soon gain a lot of experience about the changing speed of cars and the time needed to anticipate those changes. This can lead to adjusting the distance to the car in front, an important safety margin.

4.2 THE DEVELOPMENT OF RISK UNDERSTANDING

When we examine the position of a newly recruited employee at the plant, we can see that he lacks knowledge to really develop risk understanding. He lacks not only the factual knowledge but also the frame of reference to absorb new information. He knows some pieces of the puzzle, but not enough to understand whole processes and possible dangers that are involved while going about his work at the plant. Besides that, quality of his communication will also be lower, simply because he cannot yet understand the impact of the messages. Risk understanding grows year after year as a result of a deeper understanding of the processes in the plant. The frame of reference becomes mature after, for example, five to seven years, depending on the complexity of work processes.

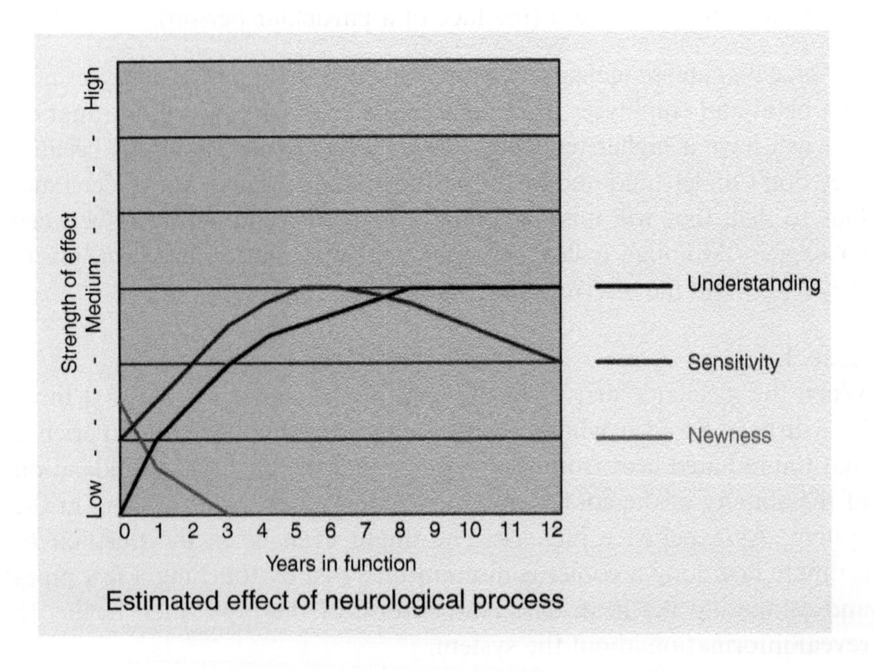

Estimated effect of neurological process

Case 1

The mechanics had a reasonably good risk understanding. They understood the process inside the cooler. The mechanic manually closed the input pipe as an extra safety measure and checked the pressure via the manual handle several times. But he never took into consideration that something might block the pipe to the valve. A better understanding of pollution inside generators or plants in general would have improved his total comprehension of the process and would have increased risk understanding. In such a case, the mechanic could have anticipated this potential problem with other checks or precautions.

Case 2

Usually driving lessons are done in reasonably good weather conditions. A young driver still has to learn how a car behaves under adverse conditions like snow or heavy showers. Once he knows how to drive in such weather, he has the potential to become a safe driver.

4.3 COMBINING NEWNESS, SENSITIVITY, AND AWARENESS

It is possible to combine the effects of the three neurological processes discussed so far. The result is an indication of the probability of safe behavior. We can see that the highest point is reached after seven years, again depending on the complexity of the processes. After that period, the weakening effect of habituation is relatively stronger than the strengthening effect of deeper insights in the work processes. If we turn the last graph upside down, we get an indication of how accident-prone an average employee can be. The conclusion is that the proneness for accidents is high in the beginning and then reduces as time passes by. After a certain period in the same function, the proneness increases again.

The estimated effects in these graphs are based on the assumption that the employee is in his first position in a plant. If the person changes jobs, then the effect of newness will decrease in the second job and the effect of sensitivity and awareness will start at a higher level. This stabilizes the final graph for safe behavior and accident proneness, although safety problems are still to be expected at the beginning or end of a period in a certain function.

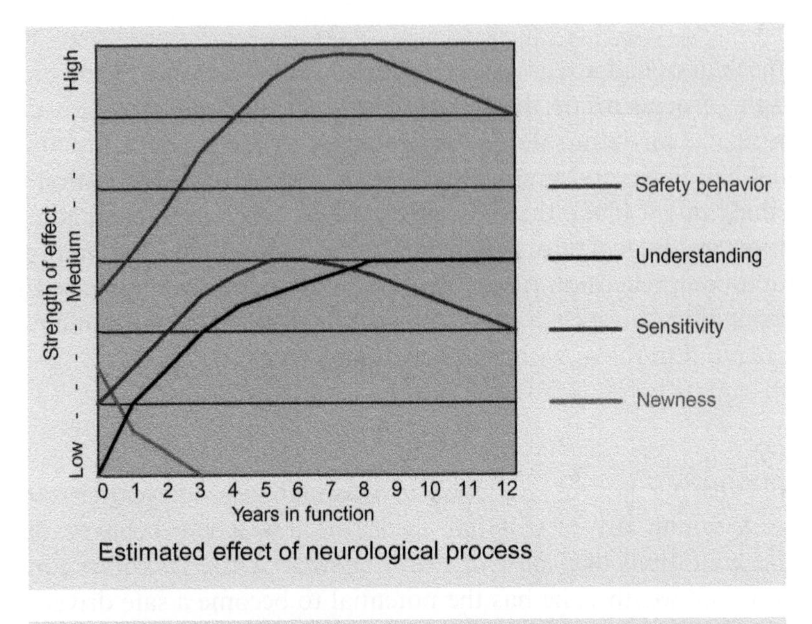

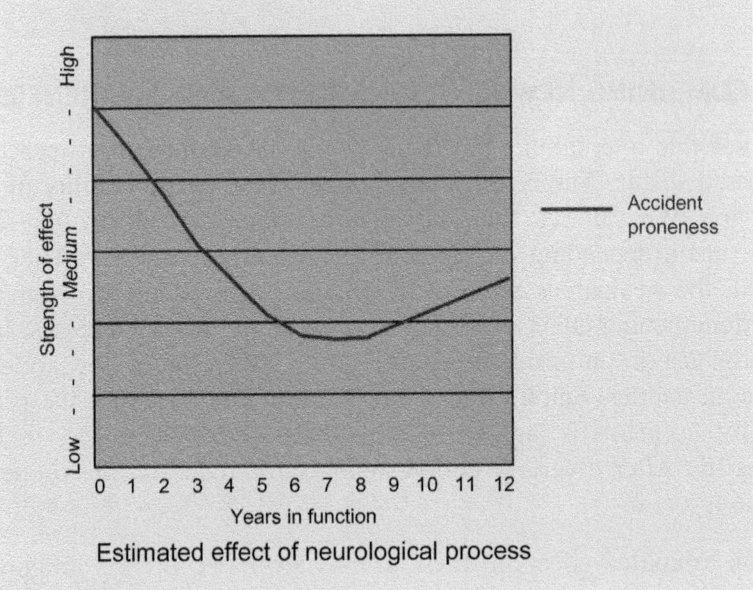

4.4 WHERE IN THE BRAIN?

Risk understanding is mostly generated by the anterior cingulate gyrus (Mobbs, 2009), an area in the center of the brain that belongs to the emotional brain. This gyrus forms the connection between the hippocampus and the prefrontal cortex. It is also the storage for the autobiographic memory. Several basic survival functions are stored here. Alexander (2011) claims that this area not only focuses on risk and safety problems in particular but on the outcome of processes in general. Attention is directed via this gyrus. It has a strong connection with the stress axis (the activation system in case we experience stress, also called the HPA, the hypothalamus-pituitary-adrenal axis). We can experience this gyrus when we have the "oops" feeling—a sudden insight that something is going wrong, sometimes combined with a shiver through the whole body. Messages from this system are handled with priority in the brain. Once the stress axis is activated, the whole system becomes prepared for a rapid reaction to a dangerous situation. The perceived danger or threat can be so strong that any possible action is estimated as being useless. In this case, the brain activity changes to a more primitive "hard-wired" pattern in which old intuitive behaviors take over. These are the moments during which we have doubts as to whether any action could still counter the danger or whether the best remaining option is to run. In this case, the brain activity changes to the basic brain (Mobbs, 2009), and this area inhibits the activity of the anterior cingulated gyrus like a hostile takeover. In the worst-case scenario, not even running is considered as an option and the person literally freezes his gestures. He feels like sticking to the floor and cannot move anymore.

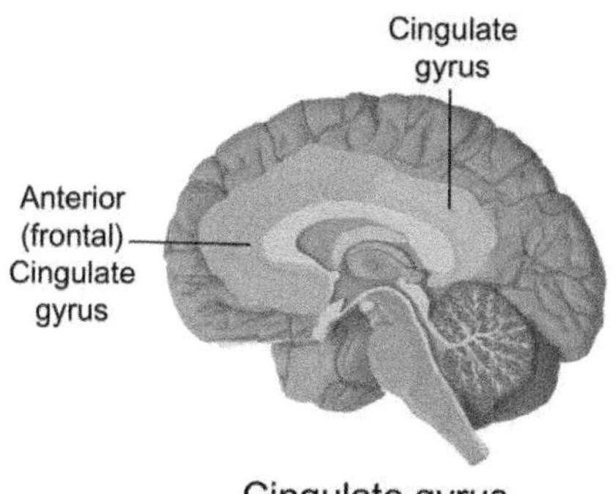

Cingulate gyrus

4.5 SUMMARY

Risk understanding is an autonomous and mostly nonconscious brain process. It scans past and present situations, extrapolates developments into the near future and checks personal plans. The quality of scanning can never be higher than the level of understanding. So the more processes are understood, the easier the brain can scan them and detect possible dangers. Time, available data, attention, and comprehending these data are relevant in developing risk understanding. Gaining expertise about the processes involved and about the impact of interventions with these work processes will support the awareness of where and how risks might appear. Employees who best understand the work process will have the highest awareness about where and which things might go wrong and how best to anticipate them. The main effort in improving risk understanding is in strengthening the frame of reference of all employees, with a special focus on safety and danger. Special attention needs to be given to the fact that each employee has a different mind-set and that information can only be absorbed if there are enough connections to the present state of knowledge.

Brain-based safety promotes all investments to create personal mastery: understanding the processes with which employees have to work. This understanding should go at least one level further than the situation demands to complete the assignment. If a team has to do a complex operation, the first level of expertise is in how each of the steps of the operation has to be done. This is the knowledge that is needed when everything goes according to plan. But as soon as something unexpected happens, a second level of expertise is needed to understand the nature of the complications.

TIPS FOR TRANSFER

Tip 1: The Toolbox Meeting

Definition: A meeting before a task starts in which all the participants discuss:

- The assignment. What has to be done and reached?
- The task. What exactly are we going to do? Which steps do we take? Which tools do we need? Who is taking which role?
- The Process. What is happening in the system we are working on? (The relevance of this is discussed in Chapter 6.)
- Team spirit. Can we give feedback on each other's safety behavior?

To structure the toolbox meeting, a checklist with the previous topics can be used. For example:

- I know exactly what the assignment is, when we have to be ready, and how the end result looks like.
- I know what I have to do in this assignment.

- I feel competent in this task and in using the tools.
- I consider the planned working method to be safe.
- I know what I can expect of the others during the task.
- In case of external contractors, I know the role division between the externals and the internals on this job.
- We have agreed on how to communicate with each other to express that it is safe to continue or to wait/stop when it is unsafe.
- I feel free to discuss safety topics while working.
- I know what the risky parts of the task are and how we can anticipate on them.

Employees from contractors have a higher level of understanding about specific tasks, but probably lack fundamental knowledge about the work process and the technical aspects of plants. Create space in every toolbox meeting for sharing information about the system they will work on, possible problems that may occur, and how to anticipate those.

Memory is partly situational: The place where you learn a fact influences the storage of that fact. If you learn and retrieve a fact on more locations, it is stored and retrieved better. So if you can organize a toolbox meeting on a location (for example, in the control room) that differs from the work location, information is stored better.

Question: Can you organize toolbox meetings, and what do you need to do to make them work?

Tip 2: Last Minute Risk Assessment (LMRA)

Definition: An LMRA is a final meeting on the spot with all the workers involved that is fully focused on assessing possible risks. A toolbox meeting should preferably precede an LMRA. If not, the checklist from the toolbox meeting could be integrated into the preparation.

During the LMRA, all the possible risks involved with doing a task are discussed once more. A checklist can be used to structure the meeting. By signing the checklist, all participants acknowledge that they understand the safety measures. In Chapter 10, we will discuss one of the principles that support an LMRA, called priming. The manager can stress the safety-first principle once more.

From a brain-based safety perspective, examples of LMRA questions are:

- Do I know what to do in case of emergency? (Am I familiar with the alarm number, emergency exit, escape route, emergency showers, fire extinguisher, assembly area, wind direction?)
- Do I understand the risks involved in the task?
- Can I estimate the possible threats connected to these risks?
- Do I know previous incidents that occurred while doing this task?
- Do I know where I can build in safety margins, take precautions so that we have room to maneuver in case a risk might happen?
- Do I know how to act in case one of the risks really appears?
- Have I taken all the needed precautions to do this task safely (workplace, location of supplies)?
- Am I wearing all needed personal protection (helmet, shoes, gloves, glasses, overalls, ear protection, and so on)?
- Am I sure the system has been rendered?
- Have I placed my personal lock so that nobody else can start the system before I have removed it?

Advice: Design the LMRA checklist in the form of a booklet that can be carried in a waistcoat pocket.

General advice: It is wise to change the questions on the LMRA card from time to time, to avoid habituation and automatic signing of the card.

Tip 3: Never a Quick Fix

The impact of the introduction of the toolbox meeting and the LMRA leaves no option open for quick fixes. Although quick fixes are usually a result of high motivation and a strong drive for performance, employees should be aware of the fact they are not fully using their brain qualities when solving problems with quick fixes. Risk understanding gets no chance in a quick fix. So even for spontaneous actions from which the organization will benefit, it is better to first make a plan, discuss it with others, let it rest for a moment, and then decide whether or not to execute the plan.

Question: How could you introduce this rule and what would be the consequences of doing this?

Tip 4: Induction

Instruct employees new in the job during the introduction period not solely on what they are supposed to do and the possible risks involved, but also on how the factory is working and why activities are done. The more understanding they have of processes in the plant, the qualities of feedstock, the transformation process, and the output, the easier they will be able to anticipate when something is not working according to plan.

Question: Is the induction organized in such a way?

Tip 5: Outsourcing

Taking into perspective that employees from contractors always lack some knowledge of the processes on which they are working, a reserved policy in outsourcing technical maintenance is preferred from a safety perspective. Although accounting rules usually press management to outsource whatever is possible, safety management flourishes with a stable crew.

To minimize this effect, it is advisable to make a deal that a more or less stable crew will serve a plant, so that the externals are able to gain a certain basic level of understanding the system. It is also wise to involve the track record of previous safety behavior in the tender of contractors.

This page intentionally left blank

CHAPTER 5

Safety Intuition

The Nonconscious Guide to Safety

In this chapter, we will discuss the nonconscious system that takes care of the major part of our safety behavior without us even knowing it. At first thought, it seems strange to imagine that the majority of our behavior stems from the nonconscious system. We see ourselves as rational, logical, and conscious beings, and compared to other animals, we indeed have more of these attributes. But after a closer look, we need to admit that we act mostly unconsciously. Just imagine the last time you had dinner with friends. When there is an interesting conversation, we taste the first bites of each new plate consciously and maybe make some compliments about the quality of the food to the cook. After that, the conversation uses all of our consciousness and we eat fully unconsciously until the plate is empty. The same happens while we are driving a car. How often are we driving on the highway and suddenly realize that we have already reached our destination? Taking into account that driving is a high-risk activity with many fatal accidents, we can no longer deny the fact that our nonconscious system takes care for most our safety.

The nonconscious takes care of all standard or previously experienced risks. It will either try to avoid them or use a safety margin to develop alternative behaviors to compensate for threatening effects. As long as we are dealing with risks included in our usual repertoire, we won't notice them. If the conscious system is active, for example, in defining the best route to a certain place, the nonconscious can give input in the decision-making process. We call this input "hunch," and define the process as intuition: knowing without knowing how we know. Unfortunately, there are a lot of biases in the nonconscious system but these can fortunately be influenced by our conscious system, at least as far as we are aware of these biases. We will discuss some of them here, especially our loss aversion and our ambivalence to deal with pain. In Chapter 4, we discussed the operational learning system that shapes behavior according to the feedback it receives. Here, we will show that the feedback system has an unfortunate influence on our safety behavior.

5.1 WHY SAFETY ALWAYS NEEDS EFFORT: UNBALANCES IN THE FEEDBACK SYSTEM OF SAFETY BEHAVIOR

The economic laws of benefits and costs are applicable to behavior. The principle is simple: If the perceived benefits are higher than the perceived costs, behavior will strengthen, and if the costs are higher, behavior will decrease and eventually stop. Unfortunately, there are two major unbalances in this economic principle, which creates a variety of reactions between people and usually decreases the tendency to express safe behavior in a modern environment. These two are:

1. Different timing of costs and rewards; costs have to be paid in advance, rewards come later
2. Differences in the visibility of consequences; costs are more visible than rewards

> Safety costs are up front; benefits will ensue later.

5.1.1 Unbalance 1: Different Timing of Costs and Rewards

The first unbalance is related to the moment of occurrence of the consequences. Safety costs are up front; benefits will ensue later. All safety costs are calculated before actual behavior starts; only a few benefits can be experienced later. People who are very sensitive to instant results have problems connecting long-term effects to previous behavior. The stronger this tendency is, the greater the probabilities that unsafe behavior is taken for granted.

5.1.2 Unbalance 2: Differences in Visibility

The second unbalance is a problem within the visibility of feedback of safety behavior. Costs of safe behavior are much more visible and tangible than rewards. This perspective makes it sometimes difficult to directly attribute the reached safety level to a specific behavior.

We always experience the safety cost but seldom realize the benefits.

These two unbalances together explain why safety management constantly needs to be supported with additional energy. The system itself is not self-supporting. We can compare it with the image of the sandglass: You have to turn the glass before all the sand reaches the ground. Once the glass has been turned around, it will work again for a while, but if you forget it, the sandglass will stop at an unexpected moment. This condition is called a hygienic condition: You only start to notice it when you are not taking care.

Case 1

The operator feels pressurized by the fact that the mechanics will be waiting for him for the duration of the two-minute walk to the car. Skipping the walk will speed up the process and is directly rewarding. At the same time, he is still ignoring the possible consequences of not checking the written assignment, but those costs are not visible yet.

Case 2

When we pass a car on a busy two-lane road, we immediately experience the increasing speed of our car, but we take for granted that we also increased the chances of a collision with a car coming from the opposite direction.

5.2 GUT FEELING, THE NONCONSCIOUS GUIDE

The brain area around the pain center has a broader function than just dealing with pain sensations. It can be regarded as a first warning system. The pain sensors lie just beneath the skin, and the intestinal tube originally also was part of the skin. The pain center in the brain has a direct link to the inner surface of this tube. This made it possible to play a role in digesting, especially in checking the quality of food. Way back in evolution, there were no refrigerators in the African savannas, and part of the consumed meat consisted of leftovers from the big hunters (lions, and so on). Quite often, the meat had become tainted, and eating tainted food is dangerous. Special pain signals from the intestinal tube were part of the danger system. As soon as the intestinal tube started sending messages that consumed food was tainted, the danger system ordered the tube to empty (vomiting and diarrhea) by contracting the muscles around it. For this reason, the danger system is closely connected to the muscles around the intestinal tube.

Long ago, our guts developed into one our first warning systems, and they still have that function.

Nowadays, if the danger system is activated, the muscles around the tube are slightly activated as a side effect of the danger awareness. This connection between the danger system and the abdomen creates the mechanism behind the "gut feeling" (Carey, 2009). The gut feeling is a warning signal that something is not alright or had better not be trusted without us knowing why. Physiologically, the gut feeling is the conscious recognition of small unconscious contractions of the muscles in the abdomen.

5.3 THE ROLE OF SMELL IN THE DANGER SYSTEM

Later in the evolution, we found out that there is an even better indicator for the quality of the food, the smell. Due to the close connection between smell and danger, the smell recognition is located next to the danger system. Smell is very important in discovering danger. Even in our language, we talk about "smelling danger." We have recently found out that smell is also very important in our estimation of reliability of other people. We will return to the topic of smell when we discuss the options of priming via smell in Chapter 9.

We can smell danger.

5.4 AMBIVALENCE TOWARD SAFETY COSTS AND THE AVOIDANCE OF UNSAFE SITUATIONS

Ambivalence originates in the construction of the brain. In modern times, the pain area is also involved in regulating psychological pains. This area becomes very active when we are being mobbed or socially excluded. Painkillers that temporarily reduce physical pain also help in the relief of depression. The pain area is even involved when we experience symbolic pains like high financial costs. It protests heavily if we are going to spend a small fortune on something. We all know the saying "it hurts in my wallet," meaning that we feel pain if we spend a lot of money. In an MRI scanner, we indeed can see the pain area becoming very active in moments when somebody makes a decision involving a certain amount of money. The higher the perceived cost, the stronger the activity in this part of the brain.

Due to the involvement of physical and symbolical pain, the pain system has to master addressing continuous dilemmas. Its task is to find a balance between avoiding pain and avoiding costs. Applying this knowledge to safety behavior, the pain system promotes a certain number of safety measures, but also protests if these extra efforts are perceived as costly. In other words, the pain system protests when a planned approach involves high-risk behavior and might damage the body or create injuries or pain. But the same pain system will also protest if a safer approach costs more in time, physical effort (precautions, procedures), or mental effort (checking plans and assumptions).

"It hurts in my wallet": Pain and costs are treated the same in the brain.

So far in this book, we have been confronted with three reasons why acting safely is not as obvious as we might think. In Chapter 2, we discussed the fact that, from an evolutionary perspective, we have developed a risk tolerance that is too high for present dangers. In this chapter, we have seen two more elements of our nonconscious system that put a burden on safety behavior. The short-term feedback we receive concerning safety behavior has a negative balance: more negative than positive feedback and hardly any visible rewards. The ambivalence of the pain center toward safety investments is the third reason. It makes each person opportunistic to a certain extent, although it will never reach the point at which it promotes dangerous actions.

5.5 THE PERCEPTION OF REASONABLE COSTS

A bias of the nonconscious system has to do with the perception of reasonable costs. The perception of cost or reward is always a comparison between an actual level of needed investment and our internal norms about how much we want to spend. This makes evaluation of safety investments personal. If several people perceive the same outcome of a specific behavior, it can either be perceived as positive, as neutral, or as negative.

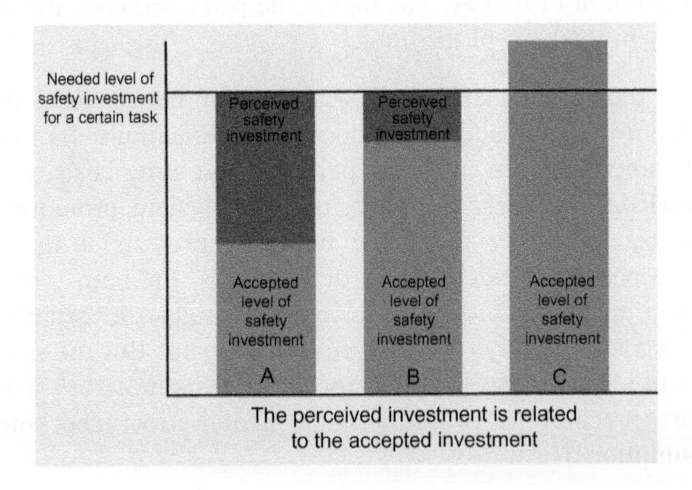

Relevant in this evaluation are expectations and needs of an observing person. If one has an urgent need for social inclusion, he will spend more effort to stay connected with his team members and will perceive connection as more valuable. The balance between costs and benefits will be perceived as more positive compared to a person who prefers solitary actions. If we transfer this principle to safety, the more safety is seen as important or essential, the less the costs for safety are perceived as a burden and the easier it will be to reach a positive economic balance for safety behavior. Deeply held attitudes function as a benchmark for evaluating costs and benefits. When three people each have a different acceptance level of safety investment for a certain task, each of them will experience the actual investment differently. To person A, who has a low acceptance level, all the safety investments will be experienced as a burden. Person C will experience the safety investments as easy, because she expected more investments than needed.

Our personal standards, our anchors, have a strong influence on the perceived costs of safety. They can, however, be influenced.

So if we can influence the level of expected and accepted investments in safety, we can also influence the perceived balance between costs and rewards and the willingness to act completely safe. Managing this level is one of the critical success factors of safety management. Kahneman (2011) introduces the concept of anchors as a point of reference concerning how much investment is appropriate for reaching a certain safety level. If we can establish high anchors (such as person C), the actual costs will be perceived as low. If it is communicated that the safety procedure will take about 15 minutes and the actual procedure takes only 10 minutes, the procedure will be perceived as quick. If the procedure was said to take only 5 minutes in a previous communication, the safety procedure will be perceived as long and costly. In the figure, the length of the green column can be regarded as an anchor. Safety anchors can be seen as part of the safety culture. We will discuss this in more detail in Chapter 11.

Case 1

If all the participants had embraced the value that no task can be started without first checking the written assignment, there would be no discussion whether or not to start. It would be obvious to wait until the assignment was checked.

Case 2

Drivers who don't become frustrated when a car in front is driving at a slow pace, because they accept that safe driving is sometimes just slow, won't try unsafe maneuvers to pass.

5.6 UNREALISTIC OPTIMISM: DENYING THE RISK PROBABILITY

Another bias in our nonconscious system leads to a reducing assessment of possible risks due to our optimism. A recent study of Sharot, Korn, & Dolan (2011) shows that we have a selective way of integrating information about risks. This leads to an asymmetry in the process of updating our beliefs of what might go wrong in the future. In the modern brain, we have a region that calculates the possible risks that might be involved in an action, based on previous experience. This region is more active in integrating positive news or estimations than in integrating negative news in the total risk assessment. So if we hear that something is safe, we incorporate this positive feedback much more easily into our database compared to negative feedback that something might be unsafe. Optimism is tied to a selective update failure and diminished neural coding of undesirable information regarding the future. We call this an attribution failure. Compensating for this unbalance with the selectivity of communication can be an option (stressing more the dangers than the safety).

> We are more sensitive to positive feedback about risks than to negative. More feedback makes us more optimistic.

Case 1

Because the mechanics are well experienced with this type of cooler, they have received a lot of feedback in the past. Habituation has reduced some anxiety, so the mechanics are quite relaxed, maybe too relaxed.

Case 2

If drivers are used to passing many times a day, they will integrate the successful passings more than the near hits. Their positive self-image as drivers who have everything under control will increase.

5.7 INTUITION: TRACES OF THE NONCONSCIOUS IN THE CONSCIOUS

The nonconscious system works without our knowing it, but does not fully operate as a black box. We receive some information out of this system, although we seldom know what this information is based on. Earlier in this chapter, we mentioned the gut feeling, produced as a warning by the danger system. This gut feeling gives us hunches about what to do or to avoid. Usually we don't know why the danger system gives this message. We experience the result (if we are able to listen to bodily signs) but not the reasoning behind it. We all have intuitive feelings from time to time, notions that pop up without us actually knowing why. People who can rely on their intuition usually make sound decisions, both in business and private life. They also anticipate possible risks.

The biggest mistake we can make concerning intuition, is that we see it as a competence independent of experience. Intuition is a very efficient and nonconscious process that combines actual perception with data mining in the several memories within our brain. We experience intuition as a hunch; in fact, it is a result of combining large amounts of stored data and drawing conclusions. Without these data, intuition cannot work. The problem is that the consciousness cannot distinguish the well-founded intuition from the poor-founded one. If the database is not filled, we cannot rely on our intuition. So we can never use our intuition on a subject with which we are not familiar. Stated differently, in new situations or with new subjects, we must forget our intuition and fully rely on facts. We cannot trust that a situation is safe; we have to check it. Without previous experience, intuition degenerates into simplification. The High Reliability Organization theory cautions for this. A lack of experience demands that we shift to the conscious system, safety awareness.

Intuition is not a competence on its own. It needs a database to rely on.

Case 1

When discussing the incident later, one of the mechanics mentioned that he had one moment of doubt when they arrived at cooler PSG-45D. He thought he heard PSG-45E, but because he was not clearly listening to the conversation between the operator and his team member in the control room, he didn't confirm it. Later, he blamed himself for not attending this little warning inside.

Case 2

Suppose a driver rents a camper for a long vacation. With a different width, weight, and engine, the camper will require another driving style. On the first day, driving the camper is so new that the conscious system is very alert and helps to adjust the existing driving patterns. After a few days, the camper starts to feel familiar and the driver might easily fall back on his previous experience and automated behaviors. The safety intuition acts as if the driver is running his car, not a camper. Experience is not yet gathered; the database is still empty. Accidents can happen more easily in this period.

5.8 WHERE IN THE BRAIN?

From the outside, the cortex looks as if it is folded, which is actually the case. It is full of curves. These are formed to increase the surface of the modern brain. The top parts are called gyri; the deep parts, sulci. On both sides of the cortex, we can see a temporal cortex, which looks like an integral part but actually hangs there like an ear warmer. The deep sulcus between the parietal cortex and the temporal cortex is called the insula, sometimes referred to as the hidden cortex because you cannot see it from the outside. It has many functions like the analysis of sounds and the integration of all stimuli (smell, taste, and touch) connected to eating food, and is important in the experience of pain.

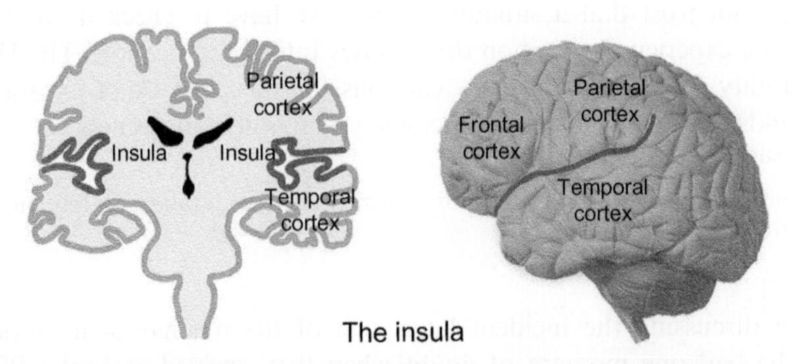

The insula

The insula connects the more exterior modern brain with the more interior emotional brain (Heuvel & Sporns, 2011). The insula scans all sensory input and all our plans on both risk and on efficiency. Inefficiency is treated in the same way as pain: Try to avoid it.

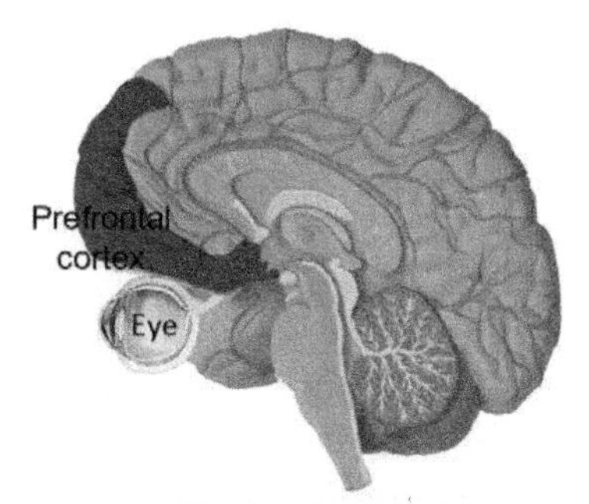

Prefrontal cortex

The part of the modern brain that gathers all feedback about a certain behavior is located in the prefrontal cortex, especially the right inferior prefrontal cortex, a few centimeters above the right eye on the side of the brain. The prefrontal cortex is famous for its reasoning center, where all the behavioral planning is done.

5.9 SUMMARY

The nonconscious system is fast, holistic, and in cases where we have previous experience, also more accurate. It takes care of our safety management without our knowing it. The system unfortunately also has some biases. The fact that the feedback we receive from safety behavior is more colored by costs than by rewards teaches us that we constantly have to invest in safety management to compensate for this. Without this investment, we become lazy and rely too much on our risk tolerance.

The origin of the love–hate relationship that some people have with safety originates in the fact that potential pain from risks and potential "pain" from precautions are handled by the same part of the brain. We want to be safe, but it shouldn't require too much effort.

How much effort we want to invest is personal and directly related to our standards. These standards can be influenced. Intuition or gut feeling can be very beneficial in detecting risk, but only in situations were we have the opportunity to gather a lot of data about safety issues. In new situations, we cannot rely on our belly; we have to check everything with our eyes, ears, and mind.

TIPS TO TRANSFER

Tip 1: Safety Investments Never Stop

The fact that we all have a brain that is programmed with a lot of risk tolerance, especially for circumstances in the 21st century, teaches us that we need a structured and continuous process to keep safety behavior on a high level. Combined with the fact that the danger center of the brain has a love–hate relationship with safety measures and that safety actions are not intrinsically rewarding, safety is not an obvious factor that will be reached automatically. Without giving continual attention to the safety process, it will lose its effectiveness. An organization will never reach a point where it can say, "Finally, we invested enough to reach complete safety."

From a safety management perspective, one could get irritated or disappointed by the fact that safety behavior has a tendency to decline, regardless of the investments that are made.

Question: What can you do to continue investing in safety improvements?

Tip 2: Bodily Sensations—Gut Feelings

By definition, we are not aware of the processes in our nonconsciousness system. There is, however, an indirect link of the conscious and the nonconscious in our bodily sensations. Although we feel these sensations in our body, our brain actually creates them. One might say they are secret messages that can guide us to some extent, if we address them. The most common recognized bodily sensation is the gut feeling. This is an innate warning system, originally created for getting rid of tainted food. Nowadays, the gut feeling is also a sign of unconsciously sensed dangers. In our society, which is dominated by the (irrational) idea that people are rational, respect for addressing bodily sensations is incorrectly low. It is advisable that management explicitly supports employees to attend to their gut feelings. During the LMRA, a simple question like "Do we have a good feeling about this job?" can help to activate the sensitivity for this kind of awareness.

Do you see opportunities to enhance this nonconscious warning system?

Advice: Always check if gut feelings are based on previous experience with similar situations.

Tip 3: Smell is a Good Warning System

We can smell angriness and anxiety from other people. There are even strong clues that we can smell cooperation. Females can pick out those men that have complementary genes, a sign of a promising combination leading to healthy offspring. Men prefer the smell of fertile women. There is a lot of research on our nonconscious ability to smell odor from other people and to relate this odor to psychological phenomena. We are much better at using our nose than we think we are. So, for example, don't try to work together with someone whose body odor is repelling to you, because the cooperation will never be a success. Unfortunately, we live in a society in which we don't give much attention to smell, and we even try to camouflage bodily odors behind a wall of deodorant and perfume. Nevertheless, if somebody thinks that someone does not smell good, please take notice of this perception and investigate it. It might lead to hidden realities.

Question: Can you recall an incident in which smell gave you a clue?

Tip 4: High Safety Anchors Create a Perception of Low Safety Investments

A safety anchor is defined as the personal standard of how much safety effort is reasonable while doing a task. If one has a high anchor, one will not easily experience safety efforts as a burden. These safety anchors can be influenced both as temporary and structural.

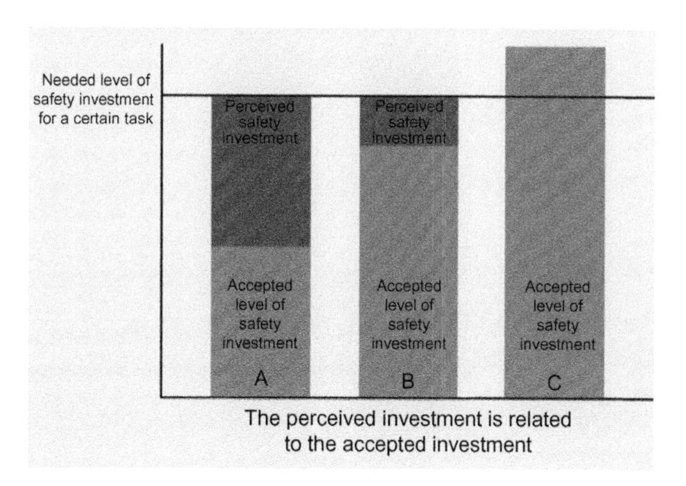

Question: Can you recognize safety anchors in your organization and can you influence them?

This page intentionally left blank

CHAPTER 6

Safety Awareness

The Conscious Guide to Safety

If we have to define the essence of human beings, we probably will start with our consciousness. We can live and at the same time reflect on this living. We can enjoy every moment, because we have a sense of time. We can learn because we know that there is more to know. But we can also be depressed because we haven't reached what we want. Consciousness is for us like water for a fish. It belongs to life and it is hard to define. Without consciousness, life has no meaning. But it's hard to understand how we can become conscious or when this happens. One of the problems is that, unlike pain or anxiety, we don't have an area in the brain that causes consciousness. We know that several areas together need to integrate their activities to create consciousness, although none of the individual areas can be held responsible for it. Consciousness can be seen as an emerging system that arises when those areas work together. Emerging systems are known in biology, physics, and mathematics. The special quality of an emerging system is that neither of the separate elements, which together create a special attribute, possesses this attribute (in this case, consciousness). For instance, color can be seen as an emerging quality. Single atoms don't possess color, but a combination of atoms into a molecule creates a total that bends the light and creates color. None of the single brain areas can create something that resembles consciousness, but a few specific areas together can.

The most appealing theory (Lamme, 2010) based on emergence is that consciousness appears if the normal nonconscious processes cannot solve a topic and start to involve many parts of the brain to work on it. This extra activity of the nonconscious is visible when standard solutions are not available in the database of previous learned experiences. We have discussed the fact that risk understanding can lead to a sudden blockade of behavior if this is regarded as dangerous. The blockade leads to conscious alarm signals, making the person aware that something is or might be wrong. Translating Lamme's theory into safety behavior, extra activity of the nonconscious system starts when there is a perceived risk that the automated programs cannot solve. This unsolved risk works like a switch to enlighten the consciousness and make us become aware in order to find a solution.

Consciousness emerges if the nonconscious cannot solve a topic.

Case 2

When a car in front of us suddenly reduces speed, the brake lights start to shine. The coordination centers notice these lights. They send parallel messages to the eyes (please watch closely) and to the anxiety centers, which check the databases. These cannot find earlier speed reductions on that particular part of the road and give a signal that this is unusual. The pain/warning center gets involved. These centers indicate that this might become serious and potentially very painful. "Why is that car reducing speed so fast?" The risk understanding confirms that extra precautions are needed. The safety margin is decreasing too fast. In the meantime, the motor area prepares a possible usage of the brake (muscle tones rise, blood pressure increases, some hormones are released to raise the sugar level in the blood, position of the foot is adjusted). As there is no confident perspective on a solution, consciousness emerges. At this moment, all the brain areas that are relevant for safety are already active and communicating with each other.

Although our knowledge about the human consciousness is still in a premature phase, we know that it is important in increasing safety behavior. In this chapter, we will expand on the topic of how we can use our awareness to enhance safety behavior and what the possible problems are in using it.

6.1 AWARENESS AND ALERTNESS

Awareness can be defined as a state of mind in which we are conscious of the here and now situation. Alertness is the conscious focus on a specific task or stimulus. Alertness always involves selection within sensory perception; some stimuli get full attention, others are ignored.

The brain consists of about 10 billion cells that have the special ability of communicating with each other. This communication is done batch wise; all the brain cells in an area send their signals at the same time. The amount of batches per second is an indicator of the speed of the communication. Awareness is strongly related to this speed, which on average is 12 Hertz[1] but can vary from 0.5 to 80 Hertz. Although there is a tendency for the speed of areas in the brain to more or less follow each other, it often happens that that some parts are more active than others, depending on which areas need to be very active. The brain uses differences in frequency to distinguish brain functions in the same way that we tune in on a certain radio frequency. Once we have tuned in on one frequency, we can focus on the messages of that particular sender.

Every activity has is own optimum speed level. In the following schedule, you can find the brain frequencies during different activities.

Brain Frequency in Hertz	Main activity	Level of awareness
0.5 to 4	Deep sleep Rapid eye movements	Absent
4 to 8	Normal sleep	Almost absent, only conditioned for specific noises, e.g., a crying child
8 to 13	Relaxed mode, comfort zone During breaks and meditation Good for absorbing information and learning (toolbox meeting)	Normal
13 to 30	Normal working mode High-risk detection (LMRA)	High
30 +	Working under high stress Incidentally also at moments of brilliant ideas	Normal to low, tunnel vision

[1]The brain speed is measured by an EEG, electroencephalography, an external monitoring of brain activity by the use of multiple electrodes placed on the scalp. The EEG mostly measures the activity of the neocortex, the modern brain, because this cortex lies closest to the surface of the brain. The EEG cannot measure any deeper activity from the emotional and basic brain.

Below 8 hertz, the brain sleeps and has hardly any awareness. The deeper the sleep, the lower the general brain activity. In the early morning, when one is getting up and at night while one is preparing for sleep, the brain frequency varies from between 8 and 13 hertz. This mode can also be reached in the daytime at relaxing moments in which we feel safe. The brain manages to operate in this modus as long as there is not too much activity needed. This is an optimal mode for learning because the brain still has room for digesting and saving new information. Because learning is one of the main purposes of a toolbox meeting, this meeting can best be held in a relaxed atmosphere; the meetings will then generate more learning and a better preparation for the task.

Every activity level has its own brain frequency.

When the need for activity rises, the speed of communication between brain cells increases. The main effect of higher brain speed is that the brain can handle situations quickly and concentrate more easily on a specific action. The brain actively gathers all relevant information that is needed for a task. It makes plans, checks options, chooses the optimal solution, and prepares for the execution of tasks. An optimum frequency for being active and detecting risks is around 20 to 25 hertz. At this speed, we have the highest awareness and an alertness that is still reasonably broad.

In a high activity mode, the possibilities to assimilate new knowledge decrease. So if people are in a hurry, for example, because they are in a tight time schedule, it is better to focus the toolbox meeting on the necessary actions. More relaxed opportunities later will probably be more suitable for learning.

With an increasing demand for action (for example, when we experience a serious threat), the general speed of the brain rises. The basic brain sends orders to the stress system. Stress can be defined as the preparation of body and brain to be ready to perform. Although stress has a negative connotation these days, it originated as a mechanism to survive. When a lion is trying to catch a zebra, the zebra probably has more stress than the lion because the zebra is running for his life while the lion is running for his meal. Within the stress system, the adrenal gland is asked to increase the production of activating hormones like adrenalin and cortisol. Both hormones have a comparable effect on the body. Adrenalin has a more long-lasting effect and can influence our body for hours; cortisol is a short-term stimulator (about 10 minutes). Due to these hormones, the blood pressures rises, the heart rate increases and more sugar is released in the blood. The body is ready for whatever might come. The alertness narrows and the focus increases. All nonrelevant stimuli are excluded from working memory to maintain maximum capacity to execute plans. Internal reasoning is on a high level. We experience the feeling of living on the edge.

6.1.1 Too Much Focus: Inattentional Blindness

While being very focused on a task, we can miss crucial information. This is called inattentional blindness (Mack, 2003; Most, 2010). A person can fail to notice a stimulus that is in plain sight. This can happen when the stimulus is unexpected but fully visible. A strong focus works like a filter and pays attention only to a few crucial stimuli directly related to a task. A new employee on a task will more easily suffer from inattentional blindness because she still needs more attention to fulfill the regular demands, which are not automated yet.

> If you need all your attention to accomplish one task, you will miss all other information.

Case 1

In this case, once the oil comes out of the pipe, everything goes so fast that there is no time for deliberation. The risk understanding increases the brain speed to a maximum within a half second. With more time, one might have discussed whether he still had manual options to close the system. Now, only the basic "run for your life" option seemed to be open.

Case 2

As soon as it is clear that the car in front has stopped and that it can hardly be avoided, the driving instructor might give some advice on what to do, but the student probably won't hear it anymore. The instructor has to act himself and cannot rely on the student anymore.

6.2 THE RELATIONSHIP BETWEEN BRAIN FREQUENCY, STRESS, AND ALERTNESS

The relationship between brain frequency and alertness can best be described as an inverted U-shape. A moderate brain frequency of 25 generates an optimal level of alertness, vigilance. At the low end, while in a comfort zone, one is not focused on anything in particular. Everything seems to be equally relevant. At the high end, one is so focused on particular cues that all the other stimuli are filtered out.

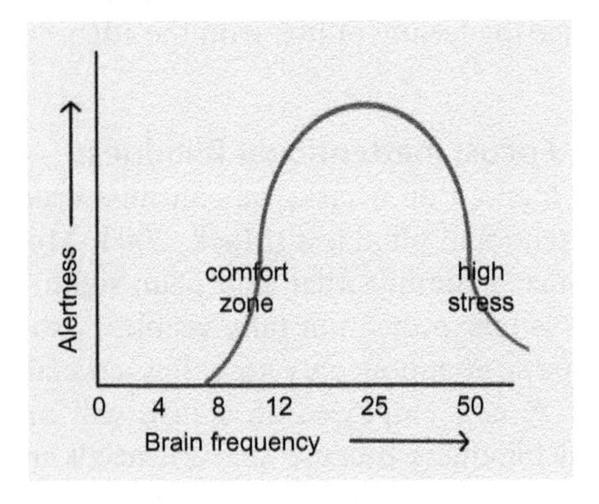

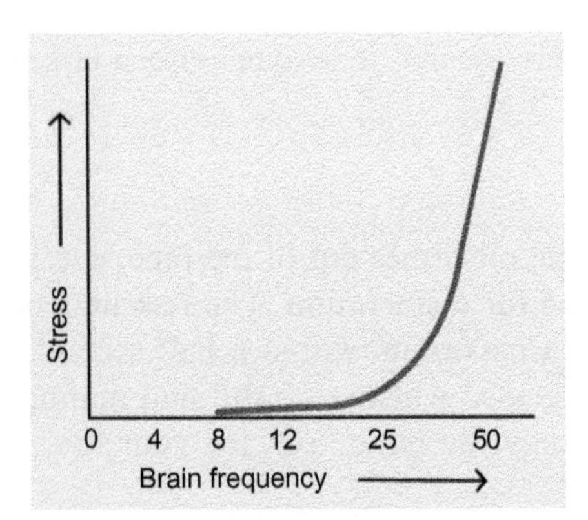

Stress has a different relationship with brain frequency. On the low end of the brain frequency, while sleeping, there is no stress, and in case we do experience stress, we don't sleep. If this is the case, the brain frequency gets no chance to drop below the 8 hertz, and we lie awake in our bed. People who suffer from burnout usually have problems falling asleep. On the high end, the stress level increases rapidly. If we experience a real threat or challenge, we suddenly can rely on a lot of stress hormones. Together, this creates a J-curve relationship between brain frequency and stress level.

6.2.1 High Stress and Safety Behavior

The origin of stress is to survive, which indicates that there is a direct relationship between stress and safety. The strange thing is that in cases of extreme stress, the consciousness has a hard time staying in control. At a brain speed of 50 hertz, the reasoning system of the modern brain can be kicked out. Although the modern brain is the most sophisticated, it will lose the internal power battle with the emotional and basic brain in cases of real danger. The conscious system is simply too slow for immediate action. The emotional and basic brains will take control and seek in their nonconscious automated repertoire for a solution to immediately reduce the perceived danger. Depending on the amount of earlier experiences and training, these immediate solutions can be very good but also catastrophic. An airplane pilot who several times a year practices in a flight simulator what to do when all the engines suddenly stop due to a cloud of dust high in the air, will probably do the right thing when he encounters such a problem because the best solution is stored in his automated behavior. The previously learned routines become leading. Highly trained emergency behavior situated in the nonconscious system can be of great benefit for avoiding primitive escape reactions. But an untrained person who needs to act suddenly might panic and do stupid things.

Once the modern brain is put aside due to a high stress level, additional information is not taken into account anymore. Advice from colleagues is no longer heard, and tunnel vision can occur (Sandi & Pinelo-Nava 2007). The irony is that people in states of high stress lose the ability to accept help, although they need it most. Under these circumstances, unreasonable actions can be taken without fully realizing the ultimate impact of the actions. Stress makes people accident prone. Later, when the stress level is back to normal again, the person involved often cannot understand his own actions anymore. When others ask for reasons behind the behavior, the person can only give fake answers, because reasoning was hardly involved in the decision-making process. Stress helps us to survive, but high stress kills those who are unprepared.

Stress creates focus, but too much stress insulates and makes you less of a team member. A tunnel vision can enhance accident-proneness.

Case 1

As soon as the oil escaped from the cooler, stress levels reached a peak within seconds. The first reaction was to fight the problem and to try to reinstall the valve. When this didn't work, the brain went into flight mode and issued to run away as quickly as possible. This flight was fully intuitive, and on familiar ground, the right route would have been chosen. In a strange environment, however, one starts to run without any plan. If there had been possible dangers in running away, this should have been trained in advance, because once high stress is reached, there is hardly any prospect for coaching.

Case 2

When the weather conditions are around freezing point, the surface of a road can be suddenly very slippery. As soon as a driver experiences that the car is losing grip and is heading for an obstacle, the stress level rises very fast. All normal reflections on how to drive are lost, and impulsive steering becomes the case. If the car is not reacting quickly due to ice on the road, this impulsive steering can lead to an overreaction, steering too fast away from the obstacle. The trip can end on the other side of the road. Only a special course for driving under extreme weather conditions can help to avoid this kind of overreaction.

6.2.2 Regulating Alertness

We have seen that each task has an optimum level of brain activity. For risk detection, the optimum would be around 25 hertz. To reach this optimum level, the brain has developed several ways of regulating the activity level. We speak of activating and deactivating systems. To some degree, each person permanently regulates his own system, but there are also circumstances that directly influence the activity of the brain. The brain can respond to external challenges and also assimilate to the activity level of the people involved in a task. Stressed people can make you more stressed; relaxed people can make you more at ease. We will now discuss several activating and deactivating systems.

External pressure increases alertness.

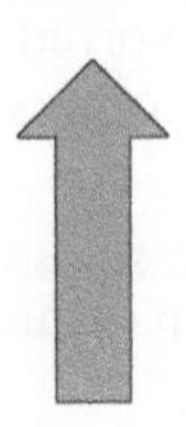

The first activating brain systems respond to increasing demands from the outside world. In general, all described processes within risk sensitivity and risk understanding can have an activating effect. Besides that, all the triggers of which are known to evoke stress, like deadlines or huge amounts of work, also increase brain activity. You can recognize this principle in people who suffer from burnout. One of the first symptoms of suffering from insomnia, poor sleep, is because the brain is activated so much that it is difficult to reach the lower activity levels needed for sleep.

Case 1

In this case, the brain of the mechanic was activated at the very moment gas and fluid escaped from the pipe. The operator activated as soon as he saw the oil squirting around, and the other mechanic as soon as he heard the operator shouting.

Case 2

The driver will be activated as soon as his risk understanding tells him that there might be a problem with a car directly in front of him.

A dynamic activation system (both increase and decrease) can be recognized in several interactions between people or with stimuli in the environment. The basic principle is called the frequency following response (Galbraith et al., 1995, Krishnan, 2004). This principle can be illustrated by the effect music can have on people. If we listen to music by Bach (with a rhythm of 60 beats per minute), we feel a different sensation than when we listen to Tiësto's trance music (with a rhythm of 180 beats per minute). These rhythms are within the possible range of the heartbeat, and our brain has a tendency to follow that frequency. Bach makes us relaxed; Tiësto invites us to dance. Chandler (2011) proves that if we speed up the process of decision making, the brain comes in a faster mode, which increases the readiness to take risks.

The frequency following response shows that, to some extent, we adapt activity levels from our environment. It is much more exciting to watch a sports game together in a stadium compared to watching that same game on TV, although we probably see the game much better on TV. Public viewing on a town square has the same effect. All the brain frequencies of the sports fans present start to align, even if you don't especially appreciate that sport and just join because you like the social activity. The same happens during a turnaround in a chemical plant, when hundreds of employees and contractors work together and try to stick to a very tight time plan. For those working, the turnaround boosts energy. Tiredness only comes when the plant is working again and everyone is allowed to feel his own tiredness. Our mirror system (see Chapter 8, important in model learning) is able to "read" the activity levels of other people. Our brain has a tendency to adapt to this level, just to belong to the group. It is an automated process.

It is possible to vary alertness with talk, self-talk, and a physical or social atmosphere.

Unfortunately, we can see the same happen when a team has an off day. Sports fans know that their favorite team, although it is the best of the league at that moment, can deliver a terrible performance during one or a few games in a row without being able to explain what is happening. In such a situation, they either drag each other into a bad modus or they have lost their ability to align due to internal conflicts, for example.

Case 1

In this case, the operator and the second mechanic immediately followed the stress level of the first mechanic. It became a collective panic.

Case 2

A very nervous student, who is about to go for his driving examination, can become more relaxed if a calm instructor is able to give him one last lesson just before the exam.

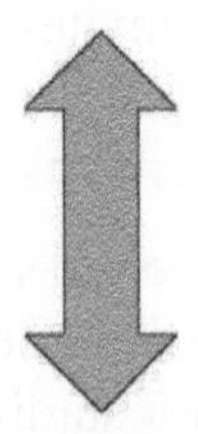

A dynamic activation system can also be recognized in the effect of feedback and self-talk. Feedback from others can either raise or decrease the level of arousal. Relaxing feedback like "don't worry" or "just take your time" reduces the arousal level, and "take care" or "we need to be very precise" raises it. Self-talk, the critical voice inside our brain, has a comparable effect. Doubting thoughts make us more anxious and raise the general activity level whereas confident, self-assuring thoughts have an opposite effect (Hatzigeorgiadis et al., 2011).

Case 1

The mechanics were probably too self-confident. They knew the task almost by heart. With less confidence, they would have insisted that the papers were on the spot before they did the LMRA.

Case 2

The feedback of the driving instructor, especially just before the exam, is of great importance for the confidence with which a student will drive during an exam. Positive feedback in the last lesson helps students to perform well during an exam.

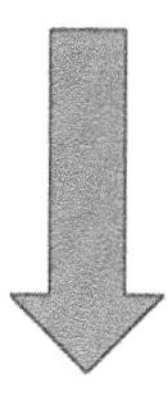

The deactivating systems start their work when we want to relax and take a break. If external pressure decreases, the brain can take a break too. In general, only the presence of others already has a relaxing effect. This is still a residue from the period in the African savannas. Hunting together is much safer. In general, a single person is more easily alerted than an accompanied one. Shared fear is half fear.

There are also active ways to deactivate the brain like mindfulness, meditation, and breathing exercises. Within the theory of the High Reliability Organization (HRO), we can see that mindfulness has become more dominant in the last 10 years (Weick, Sutcliffe, & Obstfeld, 1999). It suggests that people become much more sensitive to safety issues if they can have an open mind and can focus on the here and now situation. These are very useful instruments when the stress level is too high. Mini-breaks of 15 seconds every 15 minutes during intensive mental work can have a comparable refreshing effect on brain performance.

In cases of too much stress, relaxation methods, small breaks, and mindfulness exercises can help to reduce the stress level to a more suitable one.

Finally, some drugs like alcohol, antidepressive medication, and sleeping pills (all in the category of "this drug can influence your ability to drive a car") have a deactivating effect. They make people less worried and more relaxed. Unfortunately, these drugs disturb the dynamic system of activation and deactivation. The general level of control decreases. For this reason, they have to be considered as dangerous, and the combination between such drugs and work should be avoided.

Case 1

In this case, the people involved were very stressed until all the acute danger was eliminated. After short medical treatment, they sat together and discussed what had happened. At that moment, extreme low stress replaced high stress, a signal of emotional shock. After a few days, the people involved realized that they had been lucky. During this period, they again had a high arousal level when they talked about the incident. The next assignment on the site was again a high-arousal moment. It took a few assignments before the arousal was more or less normal again.

Case 2

Doing some breathing exercises can help a nervous student who has failed her previous driving exam to become more relaxed. A placebo pill (with the message that this drug is good to become more alert and still relaxed) can influence the self-talk and lead to a better performance.

6.2.3 Dedicating Our Alertness to a Task

Once we have optimum alertness, we still need to use it for the safe execution of a task. Even when we reach the optimum level, we only have a very limited capacity of consciousness compared to the complexity of tasks we sometimes do (Kahneman, 2011). Without the support of the nonconscious system, the conscious system would never be able to perform a safe job. The best the conscious system can do during the task is to be empty, to create a sort of naked attention, not focusing on anything in particular but being very aware of the whole situation and checking whether all things go as planned. This emptiness is sometimes referred to as mindfulness. The ideal situation is established when the nonconscious system performs the task and the conscious system just follows and uses awareness to scan everything happening.

The free workspace in our brain works best if it is dedicated to only one activity.

This perfect alignment between the two systems becomes disturbed when the conscious system gets involved with other internal or external interferences. These interferences are like obstructions that keep you from doing the best and safest job. More common disturbances are:

- Multitasking. Fast switching between two or more tasks while multitasking can kill alertness. Each task is then done with only a limited amount of attention because it takes time to load and unload the information needed in the workspace. The more complex the tasks are, the more time is needed to fully concentrate. For complex tasks, the shift takes a few minutes.
- Unfinished business. When an old task is not completed satisfactorily, the brain continues to expend energy on how to finish that task. We sometimes talk about an open gates that likes to be closed. The more open ends, the more parts of the workspace that are used for those considerations. Personal circumstances (worries about the family, children, friends, and so on) can also have the same effect.
- Worries about future events. If people start worrying about tasks that have to be done in the near future or the amount of work that never seems to reduce, these worries steal part of the workspace.
- Self-dialogue. We constantly give ourselves feedback on what we are doing, both positive ("you are doing great"), neutral ("come on, just a few... left to do"), and negative ("stupid you, you should have..."). Some people can control their internal self-dialogue very well; others almost seem to be an unwilling victim of their internal voice. Internal dialogue keeps you from giving attention to the task. Meditation and mindfulness training increase the control you have over your own self-dialogue.
- Disturbing impulses. The physical climate—for example, a climate that is too warm, too cold, or too humid, hard sounds, changing light effects, moving objects, and chitchat between colleagues—can also influence and distract attention.

All these interferences have a common effect: They temporarily steal parts of the free workspace that could be used by our alertness. By doing that, these interferences decrease the ability to work effectively and safely. To clarify this statement, we can compare the functioning of the brain with the functioning of a computer. If the actual memory of a computer is occupied by several programs that are running at the same time, the performance of the computer slows down. When we have a lot on our mind, the quality of our work also decreases. We talk about an amount of bits that is available for a task on a computer; in the brain, we talk about the amount of free cells literally available in the attention area. Part of our intelligence can be directly related to the amount of free brain cells in this area. It is one of the reasons why people start to score lower on IQ tests when they grow older. With increasing age, the amount of these free cells decreases (Daalmans, 2011).

Apart from being available, these cells also need some time to pick up the right information. They have to establish a connection with one of the databases located in many areas of brain. So if we are writing an article and we decide to answer a phone call, we have to pick up our ideas before we can start writing again. It takes some time to disconnect the cells from the information needed for the phone call and to connect again to the writing-related information. Usually, it takes a few minutes to dedicate all cells to the task again.

It takes some time before the free workspace has found all relevant information in the archives of our brain.

Summarizing, if we can focus on just one task and keep our internal and external systems quiet, we can mostly benefit from our brainpower.

Case 1

The day shift operator might have been diverted when he received the assignment to accompany the mechanics, because he suddenly had to do a series of unfamiliar tasks like meeting the mechanics, checking where to go and what to do, checking if the system had been blocked, reading the assignment, and so on. He might have received too much information to handle, and in the beginning, was still mentally busy with his previous task.

Case 2

Doing other things while driving (programming the navigator, using a cell phone, selecting music) has a very negative effect on safety behavior. Even experienced drivers start to neglect safety margins (driving too close to another car or almost hitting the side of the road on a corner) and anticipate far too slow in unexpected traffic situations (Hyman et al., 2009; Boss, 2010).

6.3 WHERE IN THE BRAIN AND THE BODY?

The actual speed in hertz is controlled by small groups of cells (the basic ganglia). These cell groups can orchestrate the speed of a brain area like a metronome. The basic ganglia are mostly situated in the basic brain, especially in the brain stem. They have very long axons spreading all over the brain, reaching almost every cortex or group of cells.

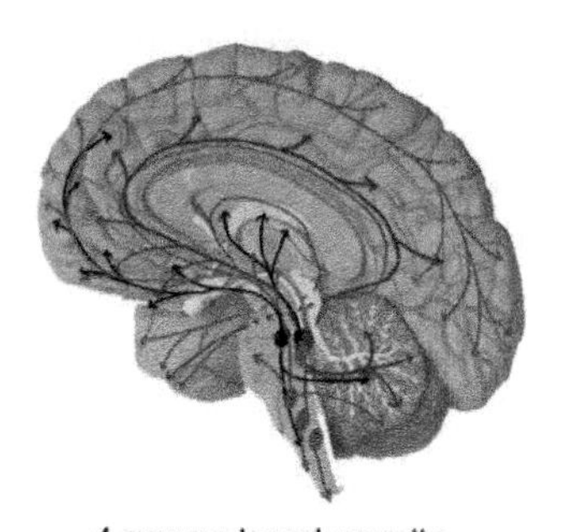

4 groups basal ganglia

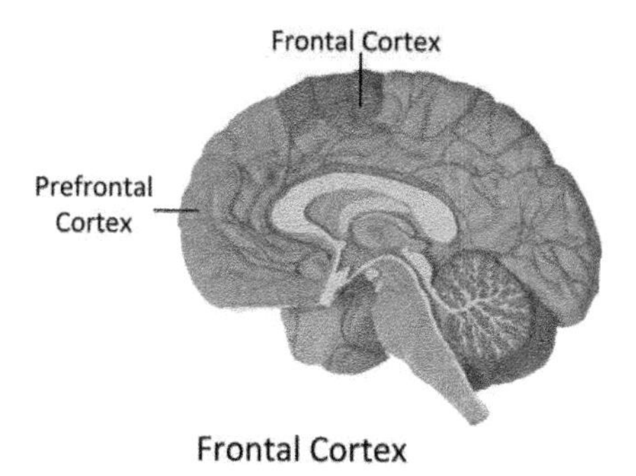

Frontal Cortex

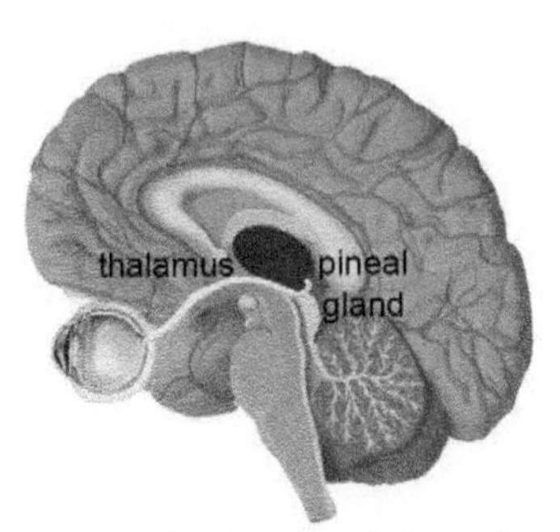

pineal gland (epiphysis)

To pinpoint and locate mental awareness is not so easy. We know that the prefrontal cortex (modern brain) just above the eyes is definitely involved, but consciousness is usually the result of the interaction between many areas. Some areas of the basic brain (thalamus and the brain stem) are absolutely necessary to create consciousness and consciousness is needed to have alertness. As long as routine tasks are done (for example, cycling), a fixed group of brain areas takes care of a task, and no conscious interference is needed. Even complex tasks, like driving a car, can be done fully nonconsciously. As soon as risks are perceived, the brain perceives this task as more dangerous and will mobilize more brain areas just to make sure that the task is done well.

Melatonin is produced by the pineal gland. This center is smaller than a pea and is located on the backside of the thalamus in the center of the brain. It is part of the basic brain. Melatonin is a hormone that prepares the body for the night. It cools down body temperature and lowers energy consumption. Receptor cells in the retina report the amount of sunlight, influencing the activity of the pineal gland. As soon as it gets dark, the pineal gland becomes more active. This gland can be fooled by tubular light with a bluish tone, faking sunlight. For those who suffer from jetlag after a night shift, a food supplement of 3 mg melatonin can compensate for the lack of natural melatonin due to staying up late. Always consult a general practitioner before using this kind of supplement.

An active brain needs an active body. The higher the frequency of the brain, the more energy it needs, the more supplies like sugar and oxygen have to be transported in the blood. Good blood pressure, rich blood, and wide veins are needed. High mental alertness will always be accompanied by higher physical readiness. The stress axis is taking care of the body that is being made ready for action.

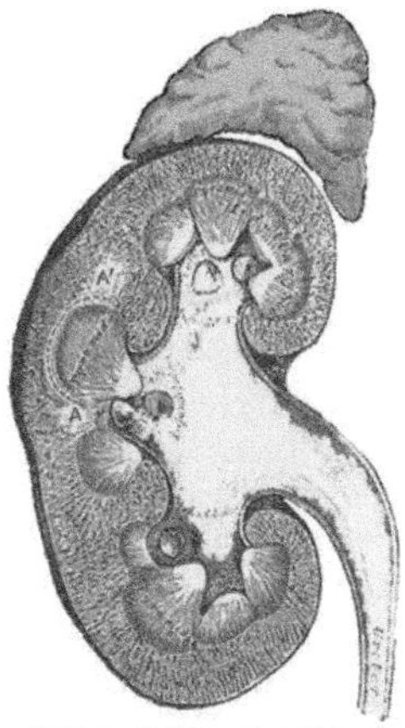

Adrenal Gland + kidney

Physical readiness is a result of an autonomous brain process that takes place without much interference from external circumstances or consciousness. The stress axis is an activity of the basic brain, that is, the hypothalamus and the brain stem (the top part of the spine in the skull). The hypothalamus sends signals to the adrenal gland that is situated on top of the kidney. This gland produces two hormones, cortisol and adrenalin. Both hormones have a comparable effect. Cortisol is mostly associated with stress and works fast and in short spurts. It shakes up the body, makes it awake, and increases the sugar level in the blood. Adrenaline covers a longer-lasting raise of general activity in the body by raising blood pressure and heart frequency, which results in a strong blood circulation. Due to this combination, muscles receive extra blood, filled with fuel. They are ready for action. The basic muscle tone increases, which can create a "speedy" sensation. Readiness requires energy. As soon as it is no longer needed, the body changes to a normal mode. The stress axis can be activated by activity in the amygdala and the hippocampus (risk sensitivity), and can be reduced by the prefrontal cortex (safety intuition) (Root et al., 2009).

6.4 SUMMARY

Consciousness makes life special but it is hard to define. Unlike other functions, consciousness is not generated by a single area, but only if a group of areas start to work together. This is probably the case when we perceive a risk that cannot be solved by our automated nonconscious programs. Consciousness is defined as being aware of the here and now situation, and alertness as the focus on special stimuli. Safety awareness has both a physical and a mental aspect that mutually influence each other. The physical aspect is related to the body's readiness to follow orders from the brain. Mental alertness varies with the frequency by which brain cells communicate with one another. A low frequency is better for learning, a high frequency for doing tasks under stress. The demands from the external world strongly influence the communication speed between brain cells. High stress is developed for acting fast while in great danger, but too much stress can reduce the impact of the consciousness because it is too slow. This can lead to automated behavior that can be dangerous when the person is not trained for that task. The brain has several mechanisms to adjust the level of alertness. Alertness uses free brain cells and can flourish in an environment that is deprived from stimuli, in a mind that is free to focus, and in a job that allows tasks to be done one by one.

TIPS FOR TRANSFER

Tip 1: Working Alone Versus in a Team

The saying "shared fear is half fear" expresses the principle that sensitivity goes down when in company. Also, anxiety levels go down when people operate together. There are many situations in which the companionship of another person is needed at work, for example, when extra hands are required, when complementary competences are needed, or when one has to do the operation and the other has to check it.

In all cases in which there is doubt as to whether two or more people are needed, consider strongly giving the task to only one person. This person will probably be more alert during the task.

Tip 2: Junior Versus Senior Employees

Juniors, defined as those new on the job, can deliver a similar result as seniors in normal circumstances. Sometimes they might even do better because they are still fresh. The biggest difference can be seen when serious problems arise and stress is involved. In such a case, a decent intuition helps to perform in a faster way. Seniors can rely on more experience, which helps them to work more intuitively. This gives them the opportunity to use their work memory more efficiently and save space for monitoring the process and its environment.

Take care that a senior is in charge when things are not going smoothly!

Tip 3: Troubles of the Mind are Thieves of Alertness

If people have a hard time on a personal level, for example, insecurity, worries, anxiety, aggression to the boss, irritation because a task is not going according to plan, keep them away from challenging tasks. In this condition, they can become a danger for themselves and for their environment.

Advise them on coaching, mindfulness training, or meditation.

Tip 4: A Free Mind is a Safe Mind—Distraction Due to Disturbing Internal Stimuli

Our mind can be disturbed by repetitive thoughts in the form of irritation, worries, doubts about competencies, and unfinished business. These thoughts occupy a part of our free working memory. As long as these thoughts keep on appearing, this part of the active memory cannot be used anymore for regular behavior. These thoughts not only disturb our well-being, but they also deregulate our unconscious behavior. Mistakes are made more easily.

If you encounter colleagues with a very occupied mind, talk with them, try to relieve their pressure if possible, and offer personal coaching, if needed.

Have you recently encountered colleagues as mentioned here?

Tip 5: Physical Healthiness Enhances Safe Behavior

The brain needs a fit body to supply it with sugar and oxygen and to remove waste materials like carbon dioxide. The quantity and quality of these supplies are directly related to physical fitness.

Can you support your colleagues in maintaining physical fitness?

Tip 6: Distraction Due to Disturbing External Stimuli

Loud sounds, moving objects, constantly changing lights, extreme temperatures, or high humidity can also lead to distraction. Create work environments that have sensory soberness. If it is not possible to adjust the environment, equip employees with tools that can reduce the impact of these stimuli like ear protection, sunglasses, or appropriate clothing.

Do you see room for improvement in this area?

Tip 7: Multitasking

Until recently, there was even a positive connotation connected to multitasking. Now we know better. A second task disturbs our conscious system, which we need for unexpected moments.

If a job consists of a series of separate tasks, do them one by one. Don't mix them if not necessary, even if it reduces the total run time. Explain that doing another job in between, regardless of how small it looks, should not be done. This creates focus, attention, and good preparation.

There were times that we thought that using the hands-free phone in cars was okay. In that period, many company cars were ordered with built-in hands-free systems. Now we know that it is better to remove them again. Phoning while driving is dangerous.

Don't allow employees to use the hands-free kit in the car, and exclude these kits from new lease contracts.

Tip 8: Alertness and Sleep

Poor sleep (insomnia or obstructive sleep apnea) has a negative effect on the quality of alertness (Kucharczyk, 2012). In this perspective, it is alarming that more than a quarter of the working population regularly suffers from sleeping disturbances due to work problems (worries, unfinished business, and technical or social problems) (NSWO, 2012). The same amount of people state they use stimulants for preventing problems that are related to a shortage of sleep (coffee, energy drinks, caffeine pills, or drugs), and 3.5 percent acknowledge that they make mistakes at work or create accidents due to a lack of sleep.

Insomnia can be work related (too much stress) but also can have medical causes like apnea. A special problem with sufferers from apnea is that they don't know they suffer from it. They just mention that they never awake rested. It should be a concern of employers, and the quality of sleep should be a fixed item in a medical checkup.

Tip 9: Alertness and Night Shifts

If we are getting tired, the frequency of the brain slows down. As soon as it gets dark, the production of a special sleep hormone (melatonin) increases, preparing the body to literally cool down. Special cells in the retina (light detectors in the eye) send messages to the brain so that it can prepare itself for the coming night. With increasing melatonin, alertness decreases.

This is the reason we become less careful during night shifts. The best remedy for this unwanted decrease in alertness is exposure to high-intensity light of a bluish color (for example, the special tubular light from Philips called Activity Light). This light mostly resembles natural sunlight. It is advisable to check the amount of lumen in the rooms where the night shift is usually working (for example, the control room) and to increase the amount of light if there is a significant difference in the amount of light during the day.

Employees who just finished the night shift (or with jetlag) can have problems falling asleep and should consult a general practitioner. The doctor might prescribe a small dose of melatonin to catch up with a normal sleeping rhythm.

This page intentionally left blank

PART 3

Influencing Safety Behavior

So far we have strongly focused on processes within a person that promote safe behavior. We have seen how people respond to risks in their environment and how they organize their behavior in such a way that they live safely. Next, we focus our attention on the role of the environment. How can actions in the outside world influence a person and his or her thoughts, plans, and behavior?

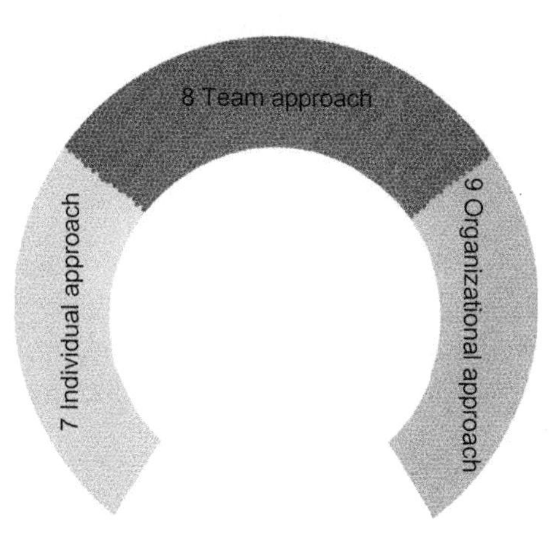

A basic assumption in this theory is that human behavior is dynamic: Although we can teach people to perceive risks, anticipate processes, and use their consciousness to find the safest solution, this gives no guarantee that people will actually behave safely. Behavior always emerges in interaction with the environment. Even if we are perfectly trained to act safely, the environment can still have such an impact that a person behaves unsafely. Changing the environment changes the behavior.

A small example can illustrate what is meant by dynamic behavior. Suppose we invite people, selected because of their friendliness, to attend a really special lecture. Once they have arrived, they are told that the response to the invitation has been overwhelming and that, because of safety regulations, only 50 percent may enter the conference hall. The ones who arrive first at a special desk will get an entrance ticket; the others, unfortunately, must return home and will only receive a copy of the slides. This message may suddenly change the behavior of these friendly people into that of hunters heading for scarce prey. Most of them will lose their friendliness on the spot and, with more or less subtle manipulation, they will try to be among the first 50 percent to arrive at the desk. The message here is that it takes more than friendly people to ensure friendly behavior; behavior is dynamic, and the environment has a strong impact on it.

If we translate this example from the concept of friendliness to that of safety, the environment has a direct influence on whether we use our learned safety behavior. Although a person may have the perfect structure in which to act safely and has acquired all the skills and knowledge to do so, he or she still can act unsafely because of external influences. This dynamic property shows the weaknesses of safety management but also gives opportunities.

It takes more than safe people to get safe behavior.

Part 3 of this book focuses on the power of external influences on both how we can foster safety behavior and what we can do to elicit this behavior once it is learned. We use three approaches: one-to-one coaching, a team approach, and an organizational approach.

In Chapter 7, we discuss how a person can be influenced on a one-to-one basis. Focusing on the role of a safety coach, called a "safety buddy," we examine the tasks, skills, and appropriate behavior for such a coach.

Chapter 8 explores how teams influence people and how individual people can influence teams. This is a more cultural approach to behavior. In order to learn what a team culture is and how it develops, we must analyze how humans are able to sense and understand other humans and how the behavior of individuals in a group can meld together into a collective, stable, and coherent society. With such an understanding, we turn next to how these processes can be influenced.

Chapter 9 discusses how an organization can influence people and what management can contribute to this process. We focus on the role of management, especially on the impact of how work is organized. Next, we examine how the organization can influence employees to participate in safety campaigns, in particular through the activation of nonconscious safety processes. The basic principle behind this influence is priming, which evokes behavior through innate or previously learned associations. A well-known form of priming is using safety posters to promote safer behavior, but priming affords a whole world of undiscovered potential. Understanding priming enables you to recognize the principles in almost all advertising, hence gaining a better understanding of why the marketing industry spends yearly billions of euros or dollars in spite of our denying that we are influenced by their campaigns.

BEHAVIOR

Before we discuss options to manipulate behavior, we must first get a better understanding of what behavior is and how behavior can change by analyzing it from a neuropsychological perspective. The actual behavior stems from the motor area in the brain that orders muscles to contract or relax so that we are able to move, speak, breath, and so on. Every group of muscles has its own coordination center in this motor area. Neuropsychology states that all our behavior is the direct result of communication among brain cells. To facilitate these communications, brain cells are connected via wires, small offshoots of one cell that physically plug into another cell, and by so doing connect these two cells so that they can communicate with each other. The illustration shows two brain cells. Each cell has a corps to receive messages, a point where it decides to send a message, a string through which the massage is sent, and at the end some connectors that transmit the message to another corps. The communication is binary; there are only two possible messages: Do more (excitation) and do less (inhibition). In this system, it is crucial which cells talk with each other, or in other words, which cells have created connections with each other. The highways and the long-distance connections are innate and for most people almost the same; they were created while we were in utero. But the many small connections are dynamic; they come and go, constantly changing during our life.

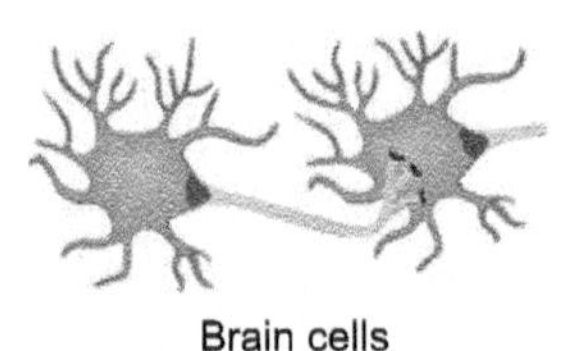
Brain cells

All behavior is the direct result of communication among brain cells.

Learning can be defined as changing the communication among brain cells. While we learn new behavior, we create new connections among brain cells. Learning happens on a physical level. Without new connections, there is no learning.

Connections can be weak or strong. In the beginning, they still are weak and can easily be destroyed. The more we show a certain behavior, the more the connections are used, the stronger they become. Behavior that is often used leads to double, triple, or even many more connections between two cells. Repetition strengthens the connections.

Learning is changing connections between brain cells.

After the learning period, the created connections stay intact as long as we keep on using them from time to time. Only if the connections are not used at all for some time will they slowly disappear because of the permanent competition for space in the skull. Biologists call this deletion "pruning." We have to realize that the brain is good in creating new connections when there is a need for communication between cells, but at the same time the brain is poor in disconnecting when this need is no longer present. Learning a new behavior is biologically much easier than unlearning an existing one. For learning, we can use our will; for unlearning, we just have to wait until the behavior disappears, and during this waiting period we can't perform that behavior anymore without reactivating it. For this reason, we can learn within seconds and establish strong connections within a day, but we also have to wait a long time before an existing link is removed again. A special problem in the deletion process is that it stops as soon as the connection is used again. That's why we easily fall back on old behavior, even if we don't want that behavior any more. Habits die slowly. If we allow old behavior to be performed again on later occasions, the weakened connections stay intact and can be easily strengthened again. So if someone wants to stop smoking or drinking and takes an "occasional" cigarette or sip from time to time, this person remains hooked.

Learning new behavior is much easier than unlearning existing one.

To enact real behavioral change, there is one alternative to the slow disappearance of old behavior: learning new behavior that is incompatible with the previous undesired one. By doing so, the new behavior blocks the existing one. So by chewing gum (with or without replacement nicotine) every time there is a craving for a cigarette, the existing patterns among brain cells are redirected and make the actual smoking behavior difficult. Because most behaviors are triggered by a specific stimulus or a previous action, it is often found effective to get rid of the trigger by following a different procedure. If a person eats too much when there are many small side dishes on the table, it is most effective to remove these tasty dishes out of sight. Likewise, if a person sometimes uses inappropriate tools, access to these tools should be reduced until a new behavior has established itself.

Fast unlearning can only be achieved by learning new behavior that's incompatible with the old one.

THE ROLE OF CONSCIOUSNESS IN BEHAVIOR

As stated before, most behavior stems from the nonconscious systems within our brain. Even most learning, like model learning, learning by classical conditioning, and learning motor skills such as cycling, happens on a nonconscious level. We can't explain with words how to ride a bicycle; a person just needs to experience it and adjust the behavior until he manages to stay on two wheels.

For some forms of learning, consciousness is involved, especially when we communicate with others. When someone explains a new policy on how to behave safely, the listener uses the conscious system to understand the message and to generate new behavior. The conscious system is able to translate messages ("Please turn this screw anticlockwise") into meaning ("Aha! She wants me to ...) and meaning into action (actually moving a tool so that a screw is turned in a certain direction). The behavior then is performed consciously, but this one-time performance does not lead to new patterns in our motor areas. In order to create these patterns, we need to repeat the behavior until the physical connections among the brain cells are strong. After several times, the behavior starts to become familiar. Within the motor area, we develop automated programs for all new elements of behavior. These new programs are integrated with the existing ones. Once they are integrated, they can be activated automatically via the nonconscious system. We only need to connect this behavioral pattern to a trigger (stimulus, idea). The learning process is finished.

Consciousness is needed for changing behavioral patterns.

CHANGING BEHAVIOR: HOW THE CONSCIOUS AND NONCONSCIOUS SYSTEMS WORK TOGETHER

But suppose we want to change an existing behavior because there is a new insight or a new safety policy. The connection between the trigger and the old behavior must be "deleted," and the new behavior has to start as soon as the trigger is active. Unlearning old behavior or deleting the connection with the trigger is difficult. Connections and patterns can only fade away slowly. Somehow we have to redirect the connection between trigger and behavior.

The easiest way to establish a new connection is via model learning. If we see others behave in an appreciated way, we are tempted to copy this behavior if the ingredients of that behavior are already part of our repertoire. If others consistently show that new behavior and we copy it, our old behavioral patterns will be disconnected from the trigger.

It is more difficult to establish new behavior all by ourselves. Habits are strong and easily triggered. First we need to learn that new behavior, which implies involvement of the conscious system. Once we have learned it, we need to take care that the trigger activates the new version and not the old one. If we behave in a nonconscious automatic way, probably the old behavior will be activated first because it still has the strongest connection. The only way to prevent this is to use our consciousness to activate and strengthen the new connection. Others, like a coach, can help us to become aware of the chosen behavioral pattern.

Case 2

Suppose a student driver has had some driving experience before his first driving lesson (for example, on his own estate or illegally on the road). During that period, the first behavior patterns are developed, and these first patterns are always strong because they don't have to compete with already existing patterns. If these patterns are not suitable for driving safely on the public road, the driving instructor will have quite a job to coach the student driver so that he performs according to best practice.

WHERE IN THE BRAIN?

Our brain is full of cells. It is estimated that approximately 25 billion of these cells are the brain cells that communicate with each other. A larger amount of other cells (like star cells) is occupied with nursing these brain cells. We can distinguish four phases in this communication process.

1. **Receiving messages.** The corps of the cell (the cell body and the small arms) receives messages from other cells. Messages have the form of chemical substances that are released near the skin of the cell. These substances are called neurotransmitters, and they create an electrical tension in the cell body. There are only two versions, plus and minus: Increase the electrical tension or reduce it. It is possible for 10,000 messages to arrive at the same time at one cell.
2. All the pluses and minuses are counted together, and the result is calculated at an edge of the cell, the hillock (the red spot in the illustration). Depending on the total electrical tension in the cell, the hillock **decides whether to send** a message. If it sends a message, this

message again is in the form of an electrical impulse that is activated in a special arm, the axon.

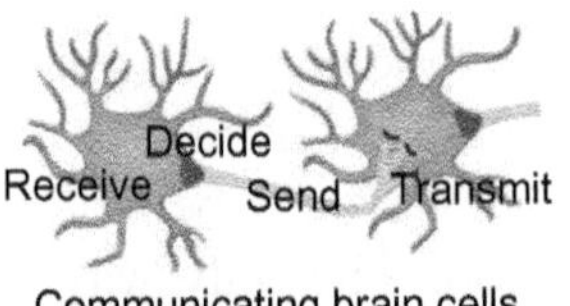

Communicating brain cells

3. The axon (yellow in the illustration) is specialized in transporting this electrical impulse. It can be **sent** to a neighbor cell or as a long-distance call to, for example, a cell at the end of the spine. At the end of the axon, it can divide itself into many small hands (blue in the illustration) called synapses. Each synapse connects to the corps of another brain cell. Sometimes more synapses connect with the same receiving cell, for example, if an extra strong message needs to be delivered. The axon can grow and create more synapses; this happens while learning. Synapses also die if they are hardly used anymore.
4. The function of the synapse is to **transmit** the message to the other brain cell by releasing a small portion of neurotransmitter.

TIPS FOR TRANSFER

Tip 1: How to Change Behavior

Once an unsafe behavior has developed, it is difficult to change it into a safer version. Our first idea for changing somebody's behavior is to explain to her why the actual behavior is not suitable and ask her to change it. We assume that supplying others with safety arguments for *why* to behave in a certain way will help them to change their behavior. Many managers are frustrated because they have explained so many times what to do and how to do it, without any effect. They don't understand why explanations don't work, and they conclude that the person involved is not willing to change her behavior.

Explanations belong to the domain of communication, which uses the conscious system to send a message. The conscious system of the receiver might or might not understand the message. A message that is understood in the conscious system unfortunately is not enough to change existing behavioral patterns in the nonconscious system. The connection between the trigger and a specific behavior is not cut or changed by a new insight.

In order to establish new behavior, we must do two things. First, we have to teach new behavior (by words and/or example) and give people the opportunity to experiment with it. Usually, we do this on a conscious level. New behavior needs practice before it can be stored in the nonconscious. Teaching new behavior is the easiest part of the transformation.

Once the behavior has been practiced and stored, a second element is needed: establishing a connection between the old trigger and the new behavior that will be stored in the nonconscious. Compare this with the way we have to shift the points of a railway connection in order for a train to take a new route. Establishing a connection between the old trigger and the new behavior can be done in several ways:

- By activating temporary consciousness in the form of a reminder made by the person himself ("When I'm there, I need to take care of ...") or by somebody else via a note ("Think about ..." or "Please take care that ...").
- By activating temporary consciousness in the form of an emotion: an association with a previous critical event (for example, your colleague broke a finger while disassembling this unit) or with an unpleasant event (for example, receiving a fine for driving too fast on the site, interacting with an angry boss)
- By using the nonconscious power of a model (for example, emulating a colleague who is doing the work in the desired way)

Question: Select one behavior of your own that you would like to change, and read the above carefully once more. Can you transfer the ideas to a practical change approach? Define what the old behavior is, what the new behavior should be, how to practice the new behavior, what the triggers are for this behavior, and ways to shift aspects of your behavior, for example, with hints, emotions or models.

Question: Now select a behavior of someone else's. What steps would be needed there?

Tip 2: Never Rely on Answers to a *why* Question

When we don't understand a behavior of another person—for example, when somebody has caused an unexpected safety incident—we tend to ask, "Why did you do this?" We expect to get an answer that explains the behavior or at least the motives behind it, and we want to use this answer to change the person, the rules, or the environment. Unfortunately, an answer to a *why* question seldom leads to a real answer; in other words, it is not wise— and is in fact even dangerous— to rely on these answers.

The first impact of the *why* question usually is a change in the power balance of a conversation. The questioner, by asking why, asks for a justification of behavior. She puts herself in a superior position and forces the other person into an inferior position. The way the question is asked, or the tone, has a strong impact on the strength of this effect on the power balance. People who feel offended easily jump into a defense mode and won't give an honest answer. The conscious thoughts might be: "I don't need to justify my behavior to you!" or "I need to give a good answer now or I will be in deep trouble."

But even if the question sounds like an open inquiry into one's personal life, the idea behind a *why* question is that all behavior is a result of will and reason, elements of which a person should be or could become conscious of. Unfortunately, contrary to general opinion, behavior usually is neither driven by a rational nor by a single will. Behavior is mostly the result of many impulses within the brain that battle for victory. Each impulse wants to be heard.

In addition, behavior is usually nonconscious and automated, and sometimes does not even aim at a certain result. A *why* question is always an intervention within the conscious system, but most behavior is nonconscious. The *why* question addresses the wrong system. For this reason, it can never lead to a proper answer. If there exists no single reason for behavior, you can never find it.

The only exception to this rule is when somebody is lying, consciously camouflaging behavior, cheating, or doing wrong things on purpose. In these cases, behavior is conscious because it wants to hide an intention. If there is a hidden agenda, the person consciously invests in keeping it below the surface. Technically, she could give the proper answer to the *why* question, but she is not interested in doing so.

In other words, the *why* question is only suitable in cases where there is a conscious violation of safety rules, and the person or group involved in this kind of situation will not be eager to give the right answer. It is not suitable for other cases in which nonconscious behavior has led to unwanted or unexpected results. As most of us expect our own behavior to be rational and intentional, we will give a socially accepted answer that sounds right. We create an explanation in hindsight and even start to believe in it ourselves. Experiments have shown that others can manipulate these hindsight beliefs in their own answers (Lamme, 2010). The conclusion is that we can never rely on an answer to the *why* question.

But what can we do to understand what has happened in case of a safety incident? First, realize that behavior is the result of three factors: our inherited structure, our learning history, and the actual environment. This leads to three different questions:

- What in our system has contributed to that behavior?
- What learning history of that person (which is partly the same as the learning history of colleagues) has contributed to that behavior?
- What elements in the environment could have elicited, stimulated, or permitted that behavior?

The best attitude to take when considering these questions is to imagine that you would have done the same if you had been in that position under those circumstances, with the same background and learning history. The more you can identify, the higher the chances that you will understand the behavior and the more realistic it becomes to define actions based on this understanding.

This page intentionally left blank

CHAPTER 7

Influencing Safety Behavior via An Individual Approach

In the next three chapters, we focus on how we can influence people to work and live more safely. We start with the individual approach. Analyzing the possible one-to-one relationships that might influence safety behavior, we can distinguish three such relationships: the line manager, the important safety models in and around the team, and finally the special role of a personal safety coach. The role of important safety models is discussed in Chapter 9, when we explore the deep impact of mirror systems in which people learn from each other. In Chapter 10, we discuss the role of the line manager and the impact of pressure on performance. In this chapter, we focus on the "safety buddy", who plays a major role in creating safe behavior patterns and in acquiring risk sensitivity and risk understanding, which can be seen as the safety foundations within each person.

7.1 WHAT IS A SAFETY BUDDY?

If a person starts a new job, he needs to be introduced to the work: the induction. This induction not only includes understanding the processes and the tasks but also how work should be done. Induction is focused on gaining competences (a mix of knowledge, skills, and experience) and attitude. The "safety buddy" can play a significant role in this induction process, both in developing competences and in acquiring the right attitude. Safety is a key component of both. Being a safety buddy is not a full-time job but should be considered as an extra task or role. A safety buddy always combines her coaching role with her regular job. She plays the role of an older sibling in the family, who partly takes care of raising a younger sibling.

A safety buddy plays an essential role in the induction of a new colleague in general and in his safety education in particular.

Case 3

The driving instructor is a perfect example of a safety buddy. Driving instructors teach a person how to drive a car, how to understand traffic, and how to become sensitive to possible dangers and anticipate them.

7.2 WHO CAN PLAY THE ROLE OF SAFETY BUDDY?

Many employees can fulfill the role of safety buddy at work. The role of the safety buddy is situational, which means that a person may play this role perfectly in one job but may still have a lot to learn in another job. Characteristics of the ideal buddy include the following:

The ideal safety buddy has seniority and a high social status within the team. He also is a safety model.

- A level of seniority in the same environment (plant, department) and task
- Competences needed to do his own job
- A knowledge of the ins and outs of the work processes
- An understanding of the safety regulations and their background
- The ability to understand and explain why the rules are the way they are and why work is organized the way it is
- Personal sensitivity to possible risks at work and to cues that might be a sign for potential disturbances in the process
- Knowledge of the risks involved in working and an understanding of the consequences if processes are temporarily organized in a different way, for example, during maintenance activities
- A work history of no longer than 10 years in this function and environment. Consider the number of 10 years as a maximum, depending on the complexity of the job. The more complex the job, the longer it takes to become a senior and the slower the process of habituation. A longer stay usually leads to a decrease in anxiety feelings as a result

of the high amount of exposure to risky stimuli, which evokes habituation. In order to be a good role model, he himself still needs to experience the anxiety that is connected to sensing certain risks, because the sharing of anxiety is crucial in the classical conditioning process.

Besides the work-related criteria, a safety buddy also has to meet two other criteria.

- A demonstrated ability to be a real champion in safety. The image of "champion in safety" is needed because a person cannot be expected to be a role model within the safety area if she is not practicing safety philosophy in all her actions. People start to disrespect a person who doesn't practice what she preaches.
- High social status in the team. The high social status is essential for a person who wants to use the power of a role model. The unconscious willingness to copy behavior from others is directly related to the attractiveness of the model. Without social status, a person has only limited modeling power.

Of course nobody fulfills all these criteria, but the closer one matches the profile, the easier it is to fulfill the role of safety buddy.

7.3 WHAT COMPETENCES ARE REQUIRED FOR A SAFETY BUDDY?

A safety buddy should acquire certain additional competences (above the regular competences of the job) before he can fulfill this role. He needs to be both a teacher and a coach. The teacher first of all understands his own additional role as safety buddy in an organization. Next, he needs to understand the theoretical background behind Brain Based Safety, especially the impact of risk sensitivity and risk understanding and the role differentiation between the conscious and nonconscious systems in behavior. He needs to understand how behavior is generated and why it is difficult to change it. Further he needs to understand that all safety behavior is dynamic. He should know how the major learning processes work because he will play an important role teaching others. He must know how he can use the principles of classical conditioning (Pavlov, offering two stimuli at the same moment) and model learning.

> The safety buddy has well developed coaching skills and understands how to act safely.

The coach is able to use his coaching skills. He understands the difference between an older brother and a parent. He can act without formal power and earns respect with his seniority. A safety buddy guides the newcomer. He can be both supportive and demanding. He can even be very straightforward if rules are about to be trespassed. He is the first one to demand that the newcomer observe the safety rules.

7.4 WHAT ARE THE ACTIVITIES OF A SAFETY BUDDY?

The safety buddy plays a crucial role in the induction of each new colleague in a job. From a Brain Based Safety perspective, a "normal" induction involves all processes in work. The safety elements of a process should be introduced as soon as the newcomer starts to understand the processes. The activity of the safety buddy starts on the first day a newcomer arrives at work. She accompanies the new colleagues on a regular basis for a period of one to two years (depending on the complexity of the processes involved) while working. Usually, when a newcomer is being taught a new tool or technique, an explanation is given of how it works, what the problems are that can be solved by it (why it is being used), what the best way is to use it, how to tell whether the tool is used well or in the wrong manner, and what can be done to avoid certain deviations. The safety buddy does the same but adds to this list where possible risks might be, what actually can go wrong, what can be done to prevent it, what the safety margins are, how to discover whether the safety margins are not sufficient anymore or that something is really going wrong, and what should be done if, although precautions have been taken, such a risk actually occurs.

> The safety buddy introduces newcomers into work processes and in all safety related aspects of these processes.

The prefect induction not only involves explaining but also regular experiencing. The newcomer not only needs to know the risks involved and how to spot them but also how to act when there is a serious threat. As explained in the previous chapter, under high stress our brain switches to a primitive mode. It neglects the participation of the conscious system and uses the fast nonconscious one. We stop thinking and we just act. Just acting without thinking is only possible if the behavior is stored in our nonconscious systems. Behavior will only be stored there if it is practiced several times and has had the opportunity to become automated. So the induction of a certain risk can only be regarded as finished when the newcomer can do the task automatically in a safe way.

7.5 THE SAFETY BUDDY AND HIS INFLUENCE ON SELF-IMAGE

Before we discuss the role of the safety buddy regarding this topic, first an introduction is required to the function of self-image in our life. In Chapter 4, it is stated that all perception is the result of a learning process. This makes perception personal: Each person perceives something unique, which can differ from others' experiences. We usually are good at estimating more objective concepts like speed, location, and size, but as soon as emotions are involved, the differences between people's perceptions grow. We do moderately well in judging certain behaviors of others, but in general we are able to judge whether or not others are acting safely. Our mirror system, which will be discussed in the next chapter, helps us do this. With respect to evolution, judging each other's safety behavior was very useful when we were hunting together. Compared to our moderately good perception of others' behavior, we are unfortunately very poor in estimating our own behavior. Most of us overestimate our own performance in many areas. This also extends to our ability to estimate the quality of our safety performance. If we ask people to rank themselves among teammates ("When ranked on safety behavior, do you belong to the top 50 percent with the safest behavior?"), 80 percent of respondents say yes and rank themselves in the top 50 percent of their own team. We call this an attribution failure (Shaver, 1970; DeJoy, 1994; Gilbert, 1995) because one falsely attributes a property (in this case, safety behavior) to oneself. This distorted self-image develops due to selective storage of information. If we behave safely, we notice it and store it as an attribute of ourselves. If we behave unsafely, we think about the reason why we do so in that particular situation and see it as an occasional deviation, not as a structural one, so we don't store it as something personal, but as situational. By this selective principle, we create a more positive image of our own behavior compared to the image others have of us. Probably the others' view is more realistic (Dobelli, 2011).

> Of all our perception, the self-image is mostly liable to attribution failures.

A too positive self-image can lead to too risky behavior without our really noticing it. It is a source of possible incidents and accidents. The only way to compensate for selective absorption of feedback into self-image is by receiving confrontational feedback from others. As people are always to some extent reserved in giving straightforward feedback, this usually doesn't happen very often. In other words, we seldom get a reality check from our outside world, especially in our work environment. It is therefore a task of the safety buddy to help the newcomer in developing a realistic self-image on safety behavior.

Case 2

We often complain about the recklessness of other drivers and while driving even use abusive words aloud as if these other drivers could hear us, but we seldom reflect on our own poor driving. If we do something stupid or drive in a risky manner, we can always think of a good reason for why we acted the way we did.

TIPS FOR TRANSFER

Tip 1: Scan your Colleagues on Suitability for the Role of Safety Buddy

For each colleague, answer the next 10 questions and count the total score.

1 = I hardly recognize this attribute.
2 = I sometimes recognize this attribute.
3 = I regularly recognize this attribute.
4 = I often recognize this attribute.
5 = I usually recognize this attribute.

1. He has a level of seniority in his work.
2. He has gained the competences needed to do his own job.
3. He has learned the ins and outs of the work processes.
4. He knows the safety regulations and their background.
5. He can understand and explain the intentions behind these rules.
6. He is personally sensitive to possible risks at work.
7. He understands the risks involved in working.
8. He still has a fresh attitude and is keen at work.
9. He has demonstrated himself to be a champion in safety.
10. He has a high social status in the team.

Scores above 35 and all scores of at least a level 2 indicate suitability for a safety buddy.

Tip 2: Tools for the Safety Buddy: A Safety Checklist

A safety buddy is an experienced employee, so she can do her induction and coaching by heart. It is, however, advisable to develop a checklist while working and to use this checklist during the induction period, just to ensure that all important elements are discussed. This checklist contains all elements that should be known in order to do a good and safe job:

- Description of the main processes at work
- Description of the critical moments in those processes, the points where the difference between normal and good performance can be made
- Description of the rules and regulations as well as the reason for each of these rules (discuss the rule in a broader audience if there is no obvious reason why to do it in the prescribed way)
- Inventory of the risky elements for safety, health, and/or environment
- Suggestions for how to anticipate possible risks and build in safety margins
- Suggestions for how to recognize, through minor details (for example, sound of the pump, heat radiation of a pipe, quality of the smoke), the status of the actual processes

In order to avoid paper monsters and more bureaucracy, it is advised that this checklist be developed in an organic way, every day a little bit, not only by the safety buddy but also by the newcomers. If the company has SHE-officers, it is wise to ask them to participate in filling in the checklist.

The checklist should be accessible all the time (maybe even through a secured Internet connection) so that it can be studied anytime.

Tip 3: Tools for the Safety Buddy: Coaching Training

There is a difference between teaching and coaching. Teaching is focused on adding information via an expert source, whereas coaching is focused on asking the right questions so that the coachee starts to think and discover. Both roles are needed in order to be a good safety buddy. Usually, teaching is much easier than coaching. A good coach withholds information for some time, although he knows it. In order to train this competence of coaching, some coaching training can strengthen the role of the safety buddy.

Tip 4: Tools for the Safety Buddy: Peer Group Meetings

To reinforce the role of safety buddy, it is good to organize safety buddy meetings from time to time. During these meetings, the safety buddies can discuss the way they are doing this task and learn from each other. These sessions will also strengthen the image of this additional role.

Tip 5: First Time Right

If you want to train a newcomer in a specific behavior, it is easiest if there are no previous experiences with this behavior. If you want to learn to play tennis, it is a nightmare for a tennis instructor to adjust your backhand if you already have played tennis some time with friends on a courtyard. Changing behavioral patterns is much more difficult then establishing them. So in the case of new behavior, organize this learning in such a way that the safety buddy is present from the first moment, and instructions can be given from the very beginning. Once the behavior is rehearsed several times and starts to become automated, the newcomer can start to practice herself.

Tip 6: Feedback on Safety Behavior

Although we all benefit when people in our environment behave safely, we are reserved in giving feedback about their safety behavior. It is a task of the safety buddy to give feedback on safety behavior in the one-on-one situation and of the manager to discuss safety topics in the team. Just imagine now what the effect would be if you asked your teammates to rank themselves by standing physically in a straight line according to their performed safety behavior. The one on the left would have the weakest safety behavior, and the one on the right, the strongest. Suppose you let them discuss why they were standing where they stood and gave them the opportunity to change position to the left or the right if they could explain why.

Question: Just for yourself, make a personal ranking based on safety behavior of your own teammates.

CHAPTER 8

Influencing Safety Behavior via a Team Approach

Now we switch our attention to the processes between people. These processes have an enormous impact on our behavior, mostly on a nonconscious level. We easily think in terms of teams, but what is a team and does such a thing exist? Relevant questions for this chapter are what the influence of a team is on the behavior of its members and whether a member can influence his team. In the theoretical framework about teams, the role of the mirror system is crucial and will be discussed intensively. Once we understand mirroring, we can also understand why changing the behavior of a group is so difficult and what options we have to do so.

8.1 WHAT MAKES A GROUP OF PEOPLE A TEAM OR A FAMILY?

It is obvious to everybody that families and teams exist. These groups act as if they are an independent unit with their own personality. Members experience themselves as being part of that group, and this membership is even seen as a part of our personality. We think we are somebody because we belong to family Y or team X).

Until recently, there was no good scientific theory that could support our layman's idea, but brain research is changing this. New neurobiological discoveries from the Italian Rizzolatti in 1996 show that we have cells or groups of cells that react to the behavior of others (Iacoboni, 2008). There is still a dispute as to whether we should talk about mirror cells or mirror systems, but this dispute does not touch the heart of the discovery. The mirror system actually mirrors an externally perceived person (Keysers, 2006). All football fans experience the mirror system when they watch their favorite team playing on TV, and the offender is close to scoring a goal. He just needs to stretch his leg and the ball will go into the goal of the opponent. While watching this moment of the game, the real fan will experience a tendency to stretch his own leg (it might even actually move) as if he could assist, through the TV, in scoring the goal. When feeding her child, a mother has a tendency to open her mouth simultaneously with that of the baby. If somebody is stuttering, we feel the tendency to complete the sentence because we are mirroring what the stutterer wants to say. We yawn when somebody else is yawning, and we feel the pain of somebody else when she is hurt. Mirrored behavior is processed in the brain as if it is our own behavior and also feels like our own behavior. For the mirror system, there is no difference between you and me. Because of this system, we can feel the same, experience the same, think the same, behave the same, and even want the same. Mirroring generates learning activities in our brain. In this way, it helps us to learn from others by just observing them: model learning.

The mirror system gives us the possibility to understand each other.

8.2 HOW DOES MIRRORING WORK?

The brain is organized in such a way that we first decide on a certain behavior (mostly nonconscious), then plan it, prepare it, perform it, internally perceive the results due to information from our muscles, externally perceive (eyes, ears) what the physical reaction of the body is, perceive how the environment reacts, and evaluate whether the behavior has the intended effect. For didactical reasons, I reduce these steps to Plan, Do, Perceive, and Evaluate. This reduction resembles the Deming circle of Plan, Do, Check, and Act, but is slightly different. By constantly following this circle, the brain can adjust the behavior in such a way that the intended behavior or the intended effect takes place. In fact, we have many of these feedback circles simultaneously all the time.

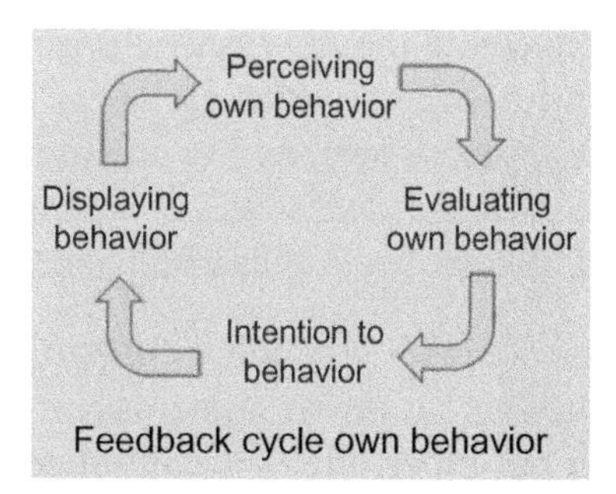

The trick of the mirror system is that the brain cells that evaluate our own behavior also can receive input from other sensory systems, in this case the observation of somebody else's behavior, that is comparable either in form or intention. So the brain cells that evaluate whether a cup is taken in the right way from the table will also be active in evaluating when somebody else takes a cup from the table. In this case, the evaluating cells are the connectors between an internal evaluating process of one's own behavior and the evaluation of a similar process performed by somebody else that arrives via external perception.

In the scheme, we add a second perception. So now we have the perception of one's own behavior and the perception of the behavior of somebody else. Both perceptions arrive at the same brain cells, and for these cells it is difficult to make a distinction between the two inputs. Because they resemble each other so much, they are processed in the same way, and in the rest of the cycle it is hardly distinguished which reaction is from which input. The perception of one's own behavior has the same status as the perception of the other's behavior. We process the intention of ourselves in the same way as the intention of the other. In other words, somebody else's intention is felt as our own intention. Even chimpanzees and gorillas have these mirror functions.

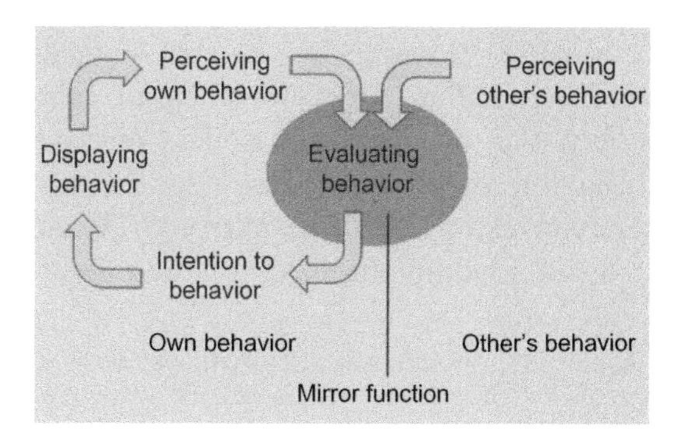

The mirror processes create the physical platform for model learning, which is a very powerful way of learning safety behavior. Mirroring mostly works on a nonconscious level and in an automatic way, but it can be controlled if we know that it exists and if we contribute via planned behavioral strategies. There are seven important aspects to mention:

1. First, mirror systems deliver a secondary function that is hosted in an existing primary function. So mirror systems can activate only behavior that is already in our repertoire. We cannot copy new behavior we don't know or yet understand. So, only after safety behavior and safety precautions are known do they become a part of our mirroring possibilities.

We can only mirror what is already in our behavioral repertoire.

2. Second, we do not simply copy the overt safety behavior but the intention behind it (Iacoboni, 2008; Schippers et al., 2009; Schippers et al., 2010). If it is clear that we believe that a behavior cannot lead to a result (for example, it is a fake behavior or a joke), we do not mirror it. So if somebody is performing a safe behavior without any intention behind it, others will not mirror it. A manager who is promoting a safety breakthrough, but who does not really believe in it herself, deep in her heart will nonconsciously show her second thoughts and will not be regarded as a role model. She might even be seen as a role model for paying lip service to safety behavior while behaving as you like.

We mirror intentions, not just overt behavior.

3. The third aspect is that we are willing to express mirrored behavior spontaneously because mirroring is a way of showing connectivity, the chameleon effect (Kühn, 2010); we integrate mirrored behavior automatically into our own repertoire. The mirror system makes parts of our behavior contagious. Please note that not only safe but also unsafe behavior can and will be mirrored. A badly performing team will destroy all good intentions of a newcomer within weeks.

Mirrored behavior is integrated automatically in our own repertoire.

4. The mirroring goes from one person to another and vice versa. If there is a group of people, mirroring processes are active between all brains in the group. Mirroring can be a great help to spread new behavioral patterns over a group, but at the same time it also stabilizes behavior. Due to these processes, new inputs of behavior can also be neutralized because the old patterns are stronger and can easily kill new routines. This is one of the main reasons why cultural change programs are so difficult to organize and are seldom effective.

Mirroring is a mutual activity.

5. Mirroring brings people together, which is intrinsically rewarding. People who mirror are perceived as more sympathetic (Kühn, 2010). Mirroring does not need any external reinforcements to strengthen the behavior; it is a self-reinforcing process. The opposite is also true: not mirroring leads to greater distance and social isolation.

Mirroring is self-reinforcing

6. The stronger the team culture, the more stable the mirroring patterns will be. A strong team with a good safety repertoire will work safely regardless of the external influences, but a strong team with a disputable safety track record cannot be changed unless the culture is weakened.
7. Role models with a high attractiveness will be mirrored more than models with a low attractiveness. Gaining sympathy from a person with a higher social status is experienced as more rewarding. In organizations, leaders are perceived as attractive models, both formal and informal ones. Peers with a high status are stronger models than peers with low status, and members of one's own team are stronger than members of other teams. This is the reason why social status is important for the effectiveness of safety buddies.

Behavior of attractive models is mirrored more

The basic question now is how we can benefit from the mirror systems to increase safe behavior in organizations. Each of the seven points below gives a clue on how to make use of this principle.

1. First we have to make sure that everybody involved knows what kinds of behaviors are needed for safety reasons. If a newcomer in a team cannot attach safety meaning to a certain behavior of a colleague, he will not copy it. Mirroring becomes much stronger, once we understand and experience a behavior.
2. We only mirror behavior if we can experience the intention behind it. Safety behavior that is exposed by a colleague just because it belongs to the protocol and not because the person believes in it, will not be copied or will hardly be copied. The only reason why a person will mirror such behavior is because she can sense the fear for punishment if she is not following the protocol. This kind of mirroring lasts as long as the fear does and will stop as soon the situation changes (for example, when alone or unable to be noticed by others).
3. Trespassing safety boundaries can be as contagious as following safety procedures. It is like a spreading fire, and it has to be extinguished. This really is a problem if all teammates trespass the same boundary.
4. The neutralizing effect of team behavior can play a role: We have to realize that new safety behavior needs to be exposed and modeled again and again before it pushes aside the old behavior. Mirroring can activate the old patterns once more, and we have seen that old patterns hardly die if they are once more used from time to time.
5. If everybody embraces the safety intentions, colleagues will act safely, just to stay part of the team.
6. If a strong team is dysfunctional on safety, sometimes the only solution is to replace a substantial number of team members or to dismantle the whole team. However, gradually changing team members one at a time keeps team behavior more intact compared to changing an entire group of team members at once. This gradual one-by-one replacement of individual members enhances the assimilation process of the individual newcomer in which he adopts the undesirable standards of the present team.
7. In change management, the colleagues with the highest attraction (sometimes called the challengers) are the first one to involve in behavioral change.

8.3 MIRRORING AND TEAM CULTURE

With the knowledge we have about mirror systems, we can explain how a team culture can develop and maintain itself. We pick up the last illustration and complete it with the internal cycle of the other person.

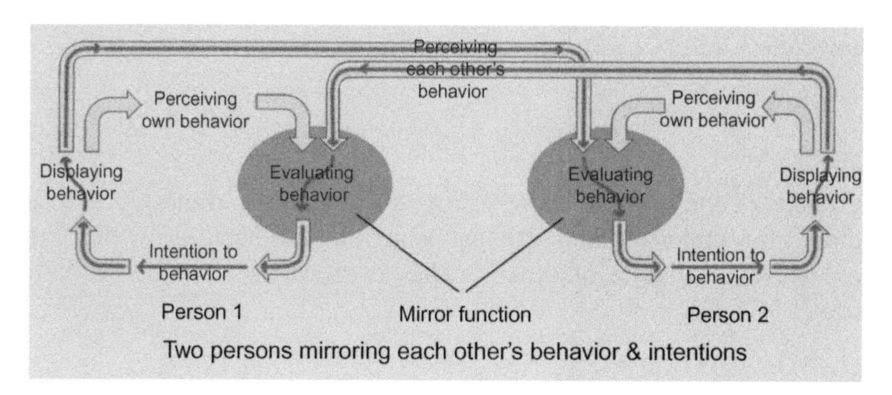

Two persons mirroring each other's behavior & intentions

Now we see two persons, each with their own feedback cycle. These cycles mutually enrich each other with mirror input from the behavior of the other person. Both persons can evaluate their own behavior, but at the same time also the behavior of the other. Research has taught us that people who imitate each other like each other more because they feel they are two of the same kind. So there is a natural tendency to copy each other's behavior, and as you can see, by doing that they also get in touch with the intentions behind those copied behaviors. Because of the mirror system, intentions mingle.

Although strictly not proven, it takes little imagination to understand that mirror systems in other areas of the brain connect not only intentions but also values and attitudes that form the foundations of overt behavior. It is very likely that mirror systems distribute values among team members. As the loops are constantly going round, there is a permanent mutual influence between team members, and from this perspective it is easier to understand why cultural change is so difficult. If one team member exposes new behavior based on new intentions but all the other team members keep on demonstrating the old behavioral repertoire, it requires a lot of resilience to resist merging with the flow of the majority and readjusting. Next in this chapter, we discuss what can be done to change these loops in a certain direction.

Case 1

In this case, we see that although the "team" of the two mechanics and the operator was new, they immediately started to develop collective behavioral norms, for example, about whether to start without having checked the written assignment, which they had forgotten and left in the car. There was also an implicit agreement to leave the cooler immediately and run away.

Case 2

As soon as the student driver starts to develop a mind-set that is suitable for understanding driving behavior of others, she will learn from the driving style of other people, even when she is not driving herself.

8.4 MIRROR SYSTEMS AND BEHAVIORAL CHANGE

We have seen that mirror systems make it possible to analyze behavior, share intentions, and probably also share values involved. At the same time, we have seen that values are constantly reconfirming themselves via interpersonal loops of that same mirror system. If we work in a cooperative way in a team of 10 persons, we constantly get feedback of the value systems of nine other colleagues. In Chapter 2, we stated that we mammals have an innate survival mechanism to stay close to the group. Because it was once too dangerous to leave the group, we have an innate tendency to stay close to each other. This closeness is also relevant in our value systems: We mostly like people who share comparable values. The basic tendency is to stick to the most common value system of the group. The stabilizing powers of a team are enormous. A fundamental question is how teams can change their value system and whether we can manipulate this change process.

As soon as one person wants to change a value because he no longer believes in the old one (let's call him a challenger, because he challenges a basic team value), he creates tension between himself and the team because he alienates himself to some extent from the general opinion. This tension is a necessary ingredient for change, but at the same time it is also unpleasant. This creates a dilemma for those who want to change a value.

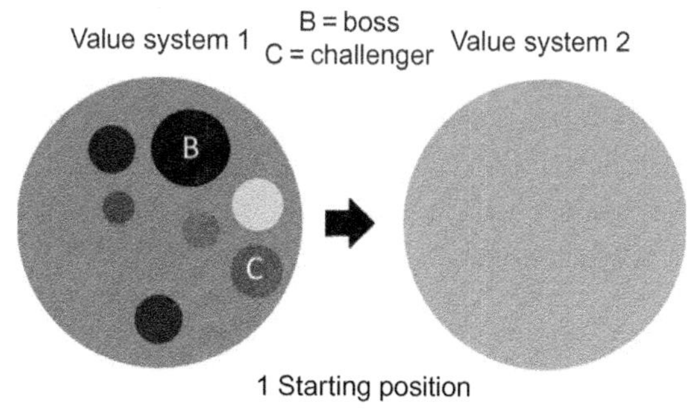

Transformation of value system

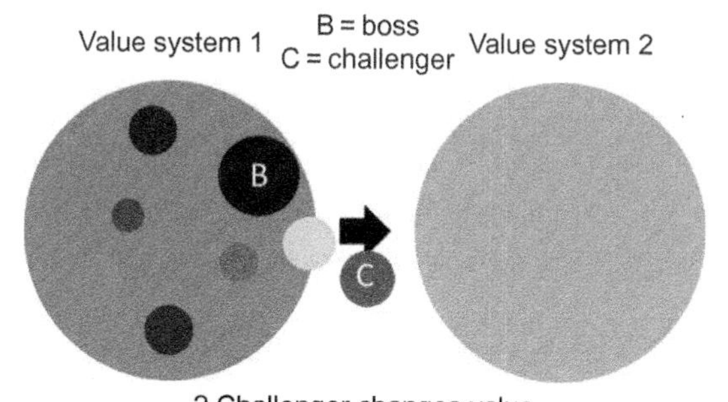

Transformation of value system

On the one hand, it takes a solitary strength to stick to one's own value if this value is different from the general standard in that team. This is difficult because a solitary position threatens the feeling of connectivity and activates anxiety about loneliness, an innate warning system not to go too far from the group.

On the other hand, we need connectivity to influence others through their empathic system. So we cannot deviate too much, or they will never follow because the distance in values has become too big.

This dilemma makes change management of values a delicate process. The challenger needs to deviate so much that he can enforce change in the rest of the team and at the same time needs to take care not to lose the contact so that others still mirror with him.

If the challenger manages to stay in contact with his team (situation 3a below) because there are some early adopters of his value system, he will still be an attractive model and others will mirror his intentions and values. A critical element in this picture is the position of the boss who, as a formal leader, also has strong power as a model. If the boss follows, he will attract most of the other team members, the early majority who helps to gain some critical mass for the new value system. At this moment, some members still lack conviction or have not yet developed the mirror abilities to sensitize themselves to the new value.

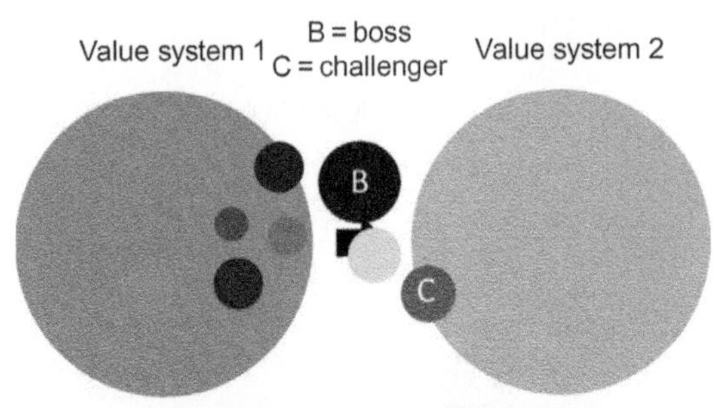

3a. Challenger attracts team members

Transformation of value system

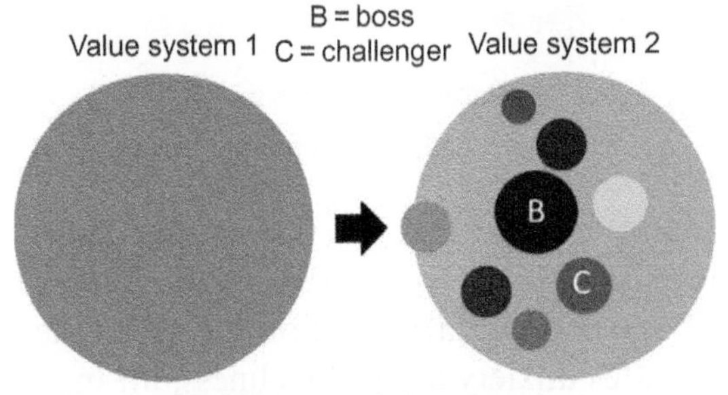

4. Challenger tempts team members

Transformation of value system

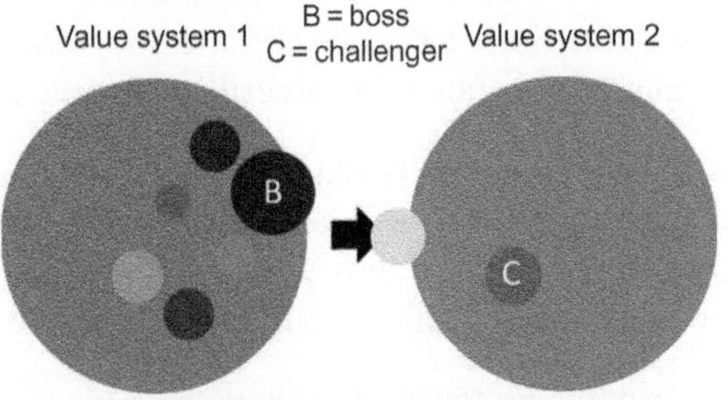

3b. Challenger loses contact team members

Transformation of value system

We have to realize that change always has a risky element, which is especially felt by the laggards. They unconsciously know that change means losing the familiar setting that they are used to and that helps them to feel safe. They don't know yet what the new value setting will bring. So they are not so eager to throw the old situation away for the new one. Only if they experience that they are drifting away from the team by not changing will anxiety about loneliness push them in the right direction.

But if the challenger loses contact with the team (situation 3b above), some followers will be confused and search for the safe haven of the old value system that guarantees the connectivity with the team members. In this case, the team loses the challenger and maybe a follower, a situation that cannot last long. As challengers have some independent qualities, the possibility that the challenger might leave the team is realistic because he does not feel at home any longer. In this example, the boss was not the challenger. This can be the case, although it usually is more effective if the boss is actually the challenger.

Summarizing, a real change occurs when the challenger is so strong that he attracts the other team members and slowly tempts them into a new value system, for example, a safer way of working.

We can also recognize this in the illustration from the two connected mirror systems. Here we see the challenger and the teammate, each with a different attitude (red and green in the illustration). If the attitudes are very different, the mirror systems won't have an influence. If a political right wing talks with a political left wing, they will only harden their points of view and create more distance. Some similarity is needed to bridge the gap between value systems.

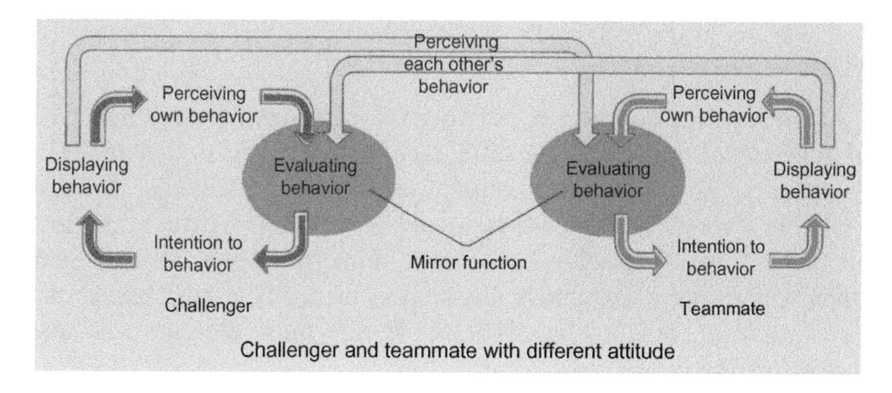

Challenger and teammate with different attitude

In this case, the green is receptive and incorporates some of the values of red. This changes green into rose. Maybe later red will also change a little more into rose, unless red is very convinced of her own point of view.

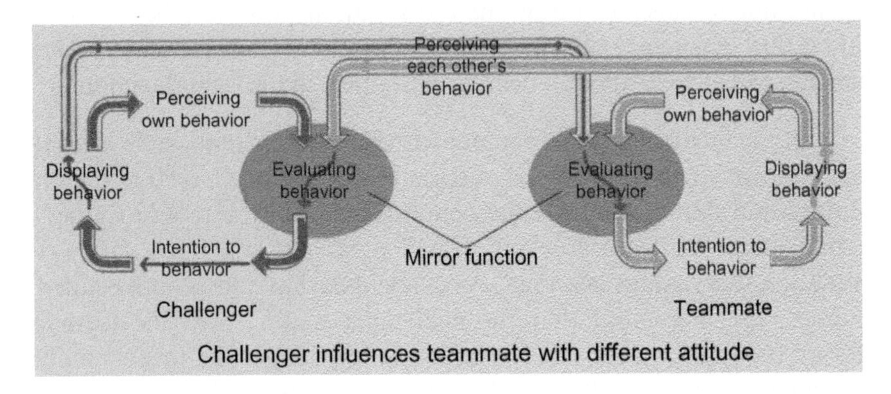

Concluding, a change of behavior and value system in a team is possible, but only gradually and with a lot of perseverance. Mirror systems work like a centrifugal force, bringing the team together in a consensus position in which deviations are only accepted to a limited degree. Knowing that we constantly have to fight unsafe tendencies, this creates an extra challenge for those who want to enhance safety.

8.5 THE SCOPE OF MIRRORING

The attractiveness of the model is an indication of the strength of mirroring processes. We mirror unless we really dislike the other person, but how strong is the impact of the model? The rule is that as long as we can identify with the model, we will adopt behavior, intentions, aims, and so on. The more we can identify, the more we will automatically mirror.

An interesting question is whether we can identify ourselves with our superiors. If there is a big difference in communication style and a big distance (physically or emotionally), we will mirror less. A bossy boss with a room one floor higher and a lot of air between his door and his desk will definitely have lower mirroring power than a team leader who is seen as almost one of the team. The tribe leader has more authority, but the father of the family more closeness, so also more mirroring power.

Twenty-five years ago, all teenage girls wanted to be dressed like Madonna; nowadays the age difference between her and teenagers is so big that she can only play the role of celeb mama. Teenage girls look for new models. Top managers often complain about an impenetrable layer of lower management that blocks all organizational innovation. This is another way of stating that they recognize that they have no mirroring power for this group. This can be caused either by a middle management layer which is not able to mirror top-down (because they are not sensitive to receive and/or transmit), or by too big a distance between the middle management and the lower management so that the middle managers aren't attractive models for the lower management (a whole layer of "Madonnas" who have forgotten that they are too old now).

> Mirroring one management level upwards is possible, two levels already creates too much a distance.

Probably, two layers of management already form a distance too big to use mirroring power. If we want to change and improve safety behavior, we need to do it with the roles of the older sibling and the father of the family: the senior colleague and the first line manager.

8.6 WHO CAN PLAY THE ROLE OF A CHALLENGER?

We must realize that we always have challengers within a team without being aware of them. It is only in behavioral change programs that we sometimes put stickers on the heads of some to increase their awareness of their role as challenger and as influencer of organizational behavior.

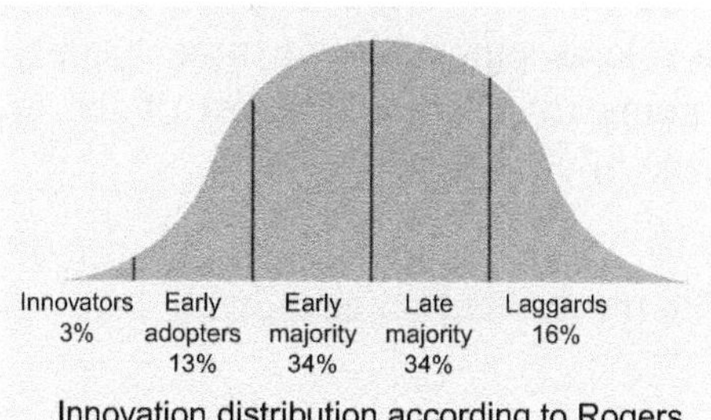

Innovation distribution according to Rogers

The innovation model from Rogers (1995) can help us differentiate the role of challenger. This model was originally developed for the marketing world but is used in many other environments. Rogers took the Gauss curve and segmented it in the usual standard deviations, which led to the distribution of 3 percent, 13 percent, 34 percent, 34 percent, 13 percent, and 3 percent. Rogers distinguished five groups of people: the innovators, early adopters, early majority, late majority, and laggards.

If we look at innovators from the perspective of mirror systems, we can only conclude that real innovators don't belong to the mainstream and that they challenge the boundaries of the usual and the accepted. They want something new and become fascinated about an idea, a product, or a way of working. They can experience this fascination because they are hardly disturbed by the general opinion. Innovators are very useful because they can be the first ones to embrace a new safety standard if they believe in it. If they accept the policy, you know you are on the right track for innovation. Unfortunately, innovators don't have the mirroring power to attract other team members. The problem is that they are so independent that they don't move around in the center of the group. They can be models, but they have problems in connecting and for this reason hardly attract others to follow them.

Innovators hardly have mirroring power.

The early adopters can play this intermediate role much better. They understand innovators and their bright ideas and can easily pick up their messages. But early adopters can also make a group enthusiastic because they are the opinion leaders that are watched (Berwick, 2003). Early adopters can give swing to group behavior. If they can be attracted to play a role as supporters for a new safety project, they are the ones that are going to sell it to the rest of the organization. They are closest to the role of challenger.

Early adopters form the flywheel for change.

Identifying innovators and early adopters is crucial if we want to use model learning through challengers to change safety behavior. Once identified, the innovators and adopters can be the first to introduce a new way of working and to sell that safety policy, respectively.

8.7 WHERE IN THE BRAIN?

The feedback loop in the brain is a clear example of a modern process, almost fully executed by the modern brain. The stages from the first idea to the final evaluation are organized in a structured way, one next to the other, from the prefrontal cortex to halfway, to the parietal cortex. There are also many small feedback loops per stage, each of which gives feedback to the previous stage.

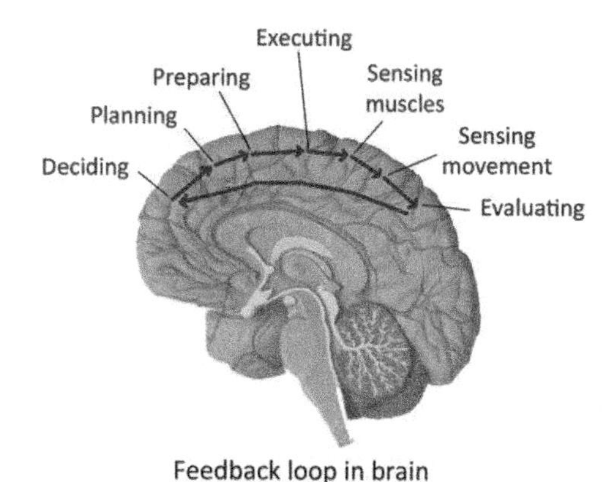

Feedback loop in brain

There is no fixed location of the mirror system. Actually, the brain has several areas with mirror systems, each located at the spot of the original function. The most important ones are located on the back of the parietal cortex, in the evaluation area. But as you can see there are many more areas, like the one on top of the frontal cortex involved in the planning of motor behavior, helping us to act synchronously. The so-called putative mirror neuron system (pMNS) consists of the premotor area at the top of the frontal cortex, the somatosensory cortex, the anterior inferior parietal lobule and the midtemporal gyrus (Schippers et al., 2010). The mirror systems in the prefrontal area help us to guess thoughts from others and to plan the same things (Kokal, Gazzola, & Keysers, 2009).

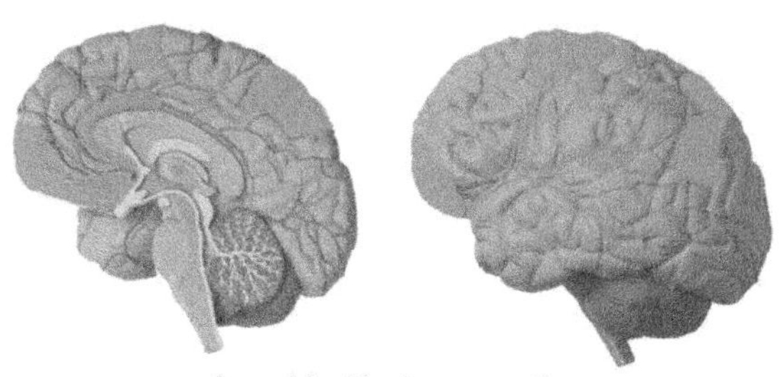

Areas rich with mirror connections

8.8 SUMMARY

Our mirror system in the brain helps us to understand and copy behavior of others. It aligns people. On a one-to-one basis, mirroring can be a strong motive in spreading safe behavior. Only known safety behavior will be mirrored from others. Good models can teach safe behavior just by demonstrating it. Fake behavior or words without acts are not copied. A team can be seen as a collection of mirroring relationships among all the members. Ten members lead to 45 different relationships. Team members mutually influence each other simultaneously via these relationships. This creates an enormous stabilizing power that results in team behavior or team culture. Cultural change means that all these connections have to change, and they also have to change at the same moment. If only a few relationships change, they will be pulled back by the other unchanged ones. The impact is that changes only can go slowly and step by step. Too strong a team can create an atmosphere in which rules are denied collectively and trespassing of safety rules become possible. Change of team behavior can only be accomplished by attracting challengers for this change. First, the most independent innovators are used to create a sense of direction. Next, the early adopters embrace the new behavior and spread it over the team.

TIPS FOR TRANSFER

Tip 1: Organizing a Team or Department Project on Safety Enhancement

If a safety project is going to be launched in the whole organization, a first step is identifying the innovators and the early majority. For each 100 employees, select the top three innovators and the (proximally) 13 early adopters.

First, discuss the new approach with the innovators. They have a good sense for newness and can give feedback on whether this approach fits to the needs of the employees and the organization.

Next invite the early adopters and ask them if they want to play a role in the implementation of the new safety approach. Train them as a group so that they identify themselves with the project.

Once they have actually acquired some of the new safety behavior, they are sent to explain it to their own teams. They can and need to express their newly acquired view in their own words within the teams. More important is that they demonstrate the vision in their actions and become a model for the rest of the team. The early adopters get a role that is slightly comparable with that of the safety buddy. The focus of the early adopters, however, is on the whole team.

Tip 2: Lip Service Versus Behavioral Commitment

Even monkeys can experience the difference between a behavior with a real intention behind it and a behavior that can be regarded as fake. Telling what you are going to do without having any real intention of doing so yourself might be believed a few times by others but will soon only be met with skepticism. So maybe there is even repair work from previous safety campaigns to be done.

If you are attempting to achieve more safety in your organization, what can you do to stay credible?

Tip 3: Absorption Time

We understand now how behavior from one person can be absorbed by a group of others, but behavioral patterns will only be strengthened if they are modeled consequent to their introduction, and for a long period of time. We also have to realize that we only mirror others if we can identify with them: We need to appreciate the model and we need to perceive the mirroring as realistic. To enhance identification processes, the difference in behavior between the model and the mirroring person shouldn't be too big.

Shaping safe behavior will be more effective if this process is approached stepwise, giving others time to absorb new behaviors one by one. Think in terms of a stepping-stone approach.

Tip 4: The Dilemma of Enforcing Rules and Regulations Versus Giving Room for Development of Attitudes on a Personal Level

Inherent to the concept of rules and regulations is the fact that these have to be followed. There is no room for a personal touch in rules. A red traffic light means stopping until it becomes green. As soon as that there is room for another interpretation (for example, red means stop unless there are no other persons or vehicles approaching), the whole system behind the rule collapses (in other words, the green light is not safe anymore). For this reason, we enforce rules: You have to follow them whatever you think of them. Zero tolerance for deviation is needed here; people have no choice.

We have seen that complete safety can only be reached if everybody embraces a set of values that puts safety first. People have to believe in it and root safety deeply into their brain (for example, in perception processes). Others can never impose this kind of conditioning; people have to embrace the values. Stated differently, people always have a choice in adopting the values or not.

So now we can experience a conflict between the approach needed for basic safety (stick to the rules) and for complete safety (embrace the rules). This is confusing for management. Is it a unique dilemma? No, we encounter it every day if we raise our children: Some rules are undisputable, and this would be sufficient if we could accompany our children for the rest of their lives. But children go out and grow up, so they have to internalize some values to guide them. As parents, we know how to combine these two contradictory pedagogic principles. How can we do that also as managers?

Tip 5: Too Strong a Team Culture

Teams can have a strong culture. In such a team, the (informal) leadership is usually well developed. The stronger the culture, the less open the team is for external influences. This can really be a problem if things are not functioning as they should—in this case, if the safety performance is not good enough.

For improvement, the strength of the culture needs to be reduced. This can be done by introducing new and strong team members and/or by replacing the informal leaders. If this option is chosen, it is advised to change everybody at the same moment. If new team members arrive one by one, they will absorb the actual team culture within a period of six weeks and hardly anything will change. If they all arrive together, they will experience mutual support as newcomers and be able to develop a countervailing cultural power within the team. In extreme cases, when the power of the team is so strong that they resist all management influence, a complete dismantling of the whole team might the only solution.

Question: Have you experienced (too) strong teams that block change?

Question: What would be your approach?

Tip 6: Optimal Size of Groups, the Extended Family

One can ask oneself what the optimal group size is to focus on behavioral change in general and safety issues in particular. The tribe size of 30 to 50 gives a good clue. We are still able to follow the behavior of others in a group of up to 50 people. Beyond that number, we lose track; we have more problems in identifying with the total group, and we start to feel lonely again. Bigger groups lead to a sense of anonymity and don't generate a sense of safety. For learning purposes, the normal department size seems to be perfect.

Tip 7: Corporate Programs for Improving Safety

It is tempting to organize companywide programs focused on cultural change. Improving safety behavior and safety management definitely belong to the agenda for such programs. The theory behind mirroring shows how difficult collective change is. A group of 100 people has almost 5000 mutually influencing mirror relationships, which all are active in mirroring behavior of each other. If we influence 100 of these relationships at the same time, the force of the 4900 other relationships still will minimize the effect. Besides that, changing behavior goes slow. Connections between brain cells don't die easily. So we need many years of constant investment in the same direction to achieve a collective behavioral change. With an average turnover of three years in management functions, it takes at least three management generations with the same ideas, values, and behavior to generate any effect. This is probably the main reason why so many corporate cultural change programs die an unknown death without concrete results.

Is this a reason never to start such programs? No, but some realism in the design of such projects is needed. Suppose you are the head of the safety department, for example, Director of Safety, Health, and Environment, and you want to enact such a project. What can you do?

First, the mirroring theory shows that the most attractive models are at the top, and their attractiveness reaches no further down than the group who can identify with them, approximately one or two management layers. Step 1 would be to do a cultural change program with the CEO and the first two management layers. If they are under the impression that their personal safety behavior is okay and that the problem is focused at the bottom of the organization, put aside for the time being a companywide project, and start local change programs like those described in Tip 6. If they see their role as being important, work with them during a period of two years and generate regular feedback. The self-attribution failure is not restricted to the bottom of the organization.

If this project is successful and leads to visible change, it can slowly spread deeper in the organization. The advised speed is to involve an additional management layer each year. Within 5 years, the whole organization is involved and within 10 years the organization can be a lot safer.

CHAPTER 9

Influencing Safety Behavior via An Organizational Approach

In this third chapter on how to influence safety behavior, we focus on the influence of the organization. First, we discuss the role of management as representative of the organization. Second, we focus on the function of rules and regulations in the organization and how we can deal with them. Finally, we focus on priming, the nonconscious influence that an organization can have via a coordinated communication of small and short messages.

9.1 THE ROLE OF MANAGEMENT

The role of management in our modern society is confusing. We are programmed in a world in which the basic community was a tribe with a tribe leader. The tribe leader acquired his position due to personal power, in early days due to a mix of physical strength and good hunting ability. The psychological contract was based on care: The tribe takes care of me, and I take care of the tribe. This care guaranteed survival and reproduction. Transferring this principle to safety: "I act safely, but

I also take care of the safety of others because their safety is essential for the prosperity of the tribe, and others take care of my safety."

Conflicting role expectations: Do managers manage care or exchange?

During the last century, we have seen a transfer from family-owned companies to money-driven companies. Suddenly the owner is an invisible stockowner and the tribe leader has become a CEO who is responsible for giving return on the invested money. This change in responsibility has also changed the psychological contract. Work is seen as the exchange of time, competencies, and money. The concept of care is replaced by a top-down responsibility from the manager for the behavior of the employee. It is expected that the boss controls the behavior of his employees and that the employees follow the commands of the boss. Transferring this principle to safety: "I act safely as far as I want to, and my boss is responsible for the safety of the environment, including the safety behavior of my colleagues."

The confusion is that we all work in the mix of two worlds: programmed as tribe member but working in the company of a stranger. Deep in our heart, we know that we have to take care of each other's safety, but we easily shift the responsibility to the boss if this is convenient for us. Management intuitively knows that leading the team as a caring family head generates the highest appreciation of team members. But management also knows that they have to sell a company policy even it is conflicting with the interest of the team. And if problems appear—for example, if trespassing of safety boundaries leads to incidents—they will be held responsible. All managers have to find their way in this confusing dilemma.

9.2 MANAGEMENT AS A MODEL

The best way to get out of this confusion is to act as a family head who himself makes no distinction between what he expects of his employees and how safely he behaves himself.

Because of his formal power, the manager has a higher attractiveness as model. He is being watched, and both his words and deeds are followed closely. A difference between doing and saying can have a very negative effect on reliability. Even monkeys can discriminate between real and faked behavior. It is very important for us to understand the behavior of others, and we are more willing to believe actions than words. A manager who actively promotes a company safety policy but behaves sometimes in an unsafe way (for example, forgetting to wear the safety belt while driving in a plant car) will be regarded with disrespect and loses his modeling power.

Behavior of the manager related to his modeling power		His personal safety behavior	
		He acts safely	He acts unsafely
His personal safety messages	He promotes safety	Ideal model Reliable	Bad model Unreliable
	He does not mention safety	Modest model	Poor model

There are two complicating factors in this story. The first factor can be described as this: Power corrupts. The more power one has, the less one feels obliged to stick to the rules. Everybody can recall top managers who trespass conventions (for example, sexual). But even from the viewpoint of employees, we can state that they regard somebody who is trespassing rules as being more powerful (Van Kleef et al., 2011). For example, when kids are playing, the strongest one can define the rules and then not stick to them.

The second factor is the discussed self-attribution failure. Managers also suffer from this failure and think that they act more safely than they in reality do. The self-attribution failure is another way of stating that we are more positive in words than in acts. Managers who suffer more from this failure are seen as less reliable, without knowing why. The only way to escape from this failure is to organize feedback so that the self-image can be adjusted to the perceived reality of others.

9.3 MANAGING STRESS

In the introduction to this book, it was stated that the combination of three basic elements determines which behavior we display. We have our evolutionary inheritance that created our basic structure (nature), our upbringing and education that taught us how to behave (nurture), and the environment that influences the actual choice for a specific behavioral option. Behavior is defined as dynamic: The readiness to act safely or to take risks is not fixed, but can change from moment to moment depending on the perceived challenges of the environment.

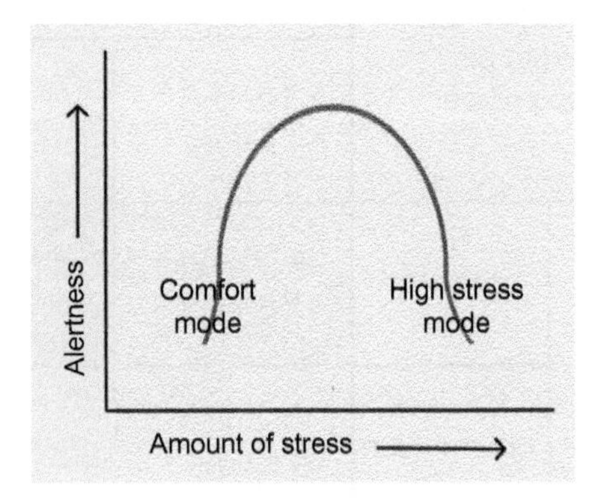

In Chapter 6, we discussed the effect of stress on alertness. The conclusion is that too low and too high a stress level both lead to insufficient safety awareness. We either fall back in a too relaxed mode and forget our safety behavior, or we get so stressed that we can't use our safety awareness anymore.

If the demands are low, the work can be done while being in a comfort zone. This leads to low alertness and a higher chance of mistakes. The biggest problems with low-stress situations is that our consciousness has so much time for daydreaming that we tend to be distracted from our task. If something unexpected happens, we still are in a relaxed mode and maybe too late to anticipate in the right way. Doing dangerous tasks while being in a comfort mode is asking for trouble. The wood sawer knows this.

High stress shuts down most internal safety systems.

When the stress rises to extreme levels, all irrelevant inputs are excluded (inattentional blindness). The modern brain makes a first estimation as to whether the actors are still in charge of the situation and how strong the threat is—"Can we still manage this or should we run?" The stronger the perception of the threat, the more the modern brain is overridden by the emotional and the basic brain. People become more intuitive and panic can result. Highly trained employees nonconsciously can take the right actions, but those who don't fully master the task become more accident prone. As panic increases, more errors are made and motor skills decline (Mobbs et al., 2009). We become clumsy, both in handling tools and in our social behavior. For highly technical tasks, the chances for a good result decrease. The brain is even preparing in case we might get hurt. Our pain sensitivity decreases (temporary analgesia) so that pain won't distracted us as much if we do get hurt. This helps to take up our last bit of energy for saving the situation (Petrovic et al., 2002). Working in a high-stress mode is asking for trouble.

Case 1

When the valve is removed, the possibility exists that a small amount of gas (nitrogen) combined with some thicker substances might come out of the pipe. As soon as oil starts squirting, there is a serious problem, an emergency. Behavior at such a moment is strongly dependent on the quality of understanding of the process. Can the valve be reinstalled? Is there a stopcock to manually block other entrances to the system? Or is there no other solution than to run? What will be the consequences of running? The higher the threat, the more the brain reverts to basic reactions. In the case of the mechanic, he gets confronted with the 3-Fs: Fighting ("Can I fight against the cause?"), fleeing ("Can I flee from the spot?") or freezing ("Is everything so overwhelming that I feel helplessly frozen and unable to act?"). These are the basic survival options. The mechanic started with fighting but soon realized that fleeing was the only real escape.

> Stress can be managed in two directions.

Stress usually has a negative connotation. Brain Based Safety states that stress is a very powerful and necessary tool to act safely. By managing the stress level, the manager can have a strong impact on the actual safety behavior of his employees. In Chapter 6, we already discussed several ways to influence the stress level. The manager can vary this level by these means:

- Making demands regarding quality and quantity of the work ("I want you to perform faster"/"Just take your time")
- Providing feedback on the way of working and the shown competences ("Come on boys, I've seen you do a better job"/"Well done")
- Focusing on safety issues ("Be careful for this task"/"Rely on the sensitivity you showed last time")
- Displaying his own level of stress ("I feel really uncomfortable about the progress"/"I estimate you are going to make it")
- Building in an emergency break, stopping the work for a few minutes, having a drink and discussing what's happening, in cases of really high stress.

Summarizing, if we want employees to work safely, we need to manage their stress level. Stress management training is a tool that can help employees to be better prepared, but the manager has an important role in regulating the amount of work to be done within a specific time frame.

9.4 MANAGING THE READINESS TO TAKE RISKS

The manager can also influence another aspect of work that is closely related to risk taking. It has been discovered that the readiness to take risks varies with the perception of the possible outcome. The psychological law goes like this: We play safe if we can keep what we have, and we take a risk if we can reduce a loss (Thaler, 1980, Tiemeijer, 2011). If outcome is perceived as a loss, we are more eager to take risks. If we perceive outcome as a gain, we prefer to play safer. In Chapter 2, we discussed the concept of loss aversion. This principle is directly related to avoiding losses. During a game, we reduce risk once we are winning, and we increase risk once we are losing. A gambler who is losing is more willing to take risks and even trespass safety boundaries if he can avoid a possible loss. It is the mechanism behind huge gambling debts.

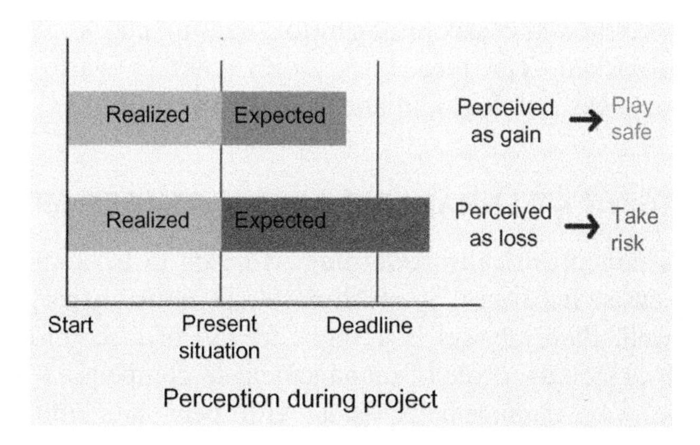

This law creates a special problem in all cases where targets are involved. As long as we haven't yet reached the target, the difference between the already achieved level and the target is perceived as a loss. When the deadline is near and we still have some activities for a final task, we become more willing to take risks. According to this rule, unsafe behavior is more likely to occur just before the deadline of a project.

If we are about to miss a crucial deadline, we can easily be seduced to take an extra risk, just to finish in time.

Crucial to this process is how the person involved perceives the task: Can it lead to a possible loss or to a possible gain?

The manager needs to act if he discovers that employees are stretching themselves so heavily that their willingness to take risks increases. He can change the deadline by giving extra time or manage additional support and supplies so that the work can be done faster. It can be argued that it is a sign of bad management to accept trespassing of deadlines, but if employees are really stressed, they can't maintain their level of performance. They will more easily make mistakes, which can lead to even further delay. Taking away the direct trigger for stress is the best guarantee that the work will be done both safely and quickly.

Case 1

In this case, preparation before starting the task took significantly longer then expected. The mechanics first had to wait for the usual operator, and then the second operator had to check whether the cooler was blocked. The mechanics didn't want to waste any more time; the clock was running, and another job was waiting.

9.5 MANAGING AN ENHANCING SAFETY ATMOSPHERE

As safety is not intrinsically rewarding, it needs to be managed on a permanent basis. Because it is much easier to notice aspects that are not going well, there always is a slight feeling of discomfort around this topic. It is seen as a role of management to compensate this atmosphere. The way management deals with near hits and incidents strongly influences the perception of the employees.

Weick, Sutcliffe, and Obstfeld (2005) introduced the concept of sense-making as a way of turning circumstances into a situation that is comprehended explicitly in words and that serves as a springboard into action (Taylor & Van Every, 2000). The main idea is to focus on the impact of how we deal with circumstances and what kinds of words we use while dealing with it. The best way to explain this is by giving two options on how to approach a safety incident caused by an operator.

In the first approach, we focus on the operator and ask him why he acted the way he did. Apart from the fact that we never get a real answer on a *why* question (see Tip for transfer in the introduction of Part 3), by doing so the discussion can easily be interpreted as blaming a decision that caused the incident. It is only a small step from blaming the decision making to blaming the one who made the decision, the operator. Once the sense of blame is in the room, there is no fun anymore in researching what went wrong and how this can be prevented next time. If the operator later shares his perception with his own team, the willingness to report future incidents will decrease.

The alternative is to analyze that same incident by focusing on the task and the context in such a way that what happened starts to make sense in a broader perspective. As soon as we can analyze the incident without analyzing the actor, but focusing instead on the act, it creates an opportunity to personally identify with what happened. We might even understand the incident in such a way that we realize that we could well have acted the same way in such a situation. At that moment, learning starts because we fully experience the act in that specific situation, which gives us the opportunity to reframe what happened in such a way that we get a deeper understanding. This understanding can lead to new insights in how to organize that situation in such a way that nobody is tempted anymore to behave in that way. The theory states that as soon as an interpretation makes sense for us, it also creates sense in how to deal with these kinds of situations. Sensemaking is creating a new perspective together that gives new meaning to the same facts.

Only by really understanding the behavior that led to an incident can we use it for improving safety.

It can be said that this sounds good as long as there is an unforeseen mistake. But what happens if the incident was the result of consciously trespassing safety rules? Even then, management can try to understand why an employee trespassed that rule. There can be many reasons for this behavior, and it is very helpful to find out what really happened in the mind of that person. Everybody wants to leave work in the same shape in which he or she arrived; nobody deliberately wants to get hurt. So there is always a pattern, however awkward it might be. Understanding these patterns is much more effective than giving fines or penalties.

This example shows that the way in which we ascribe words to certain situations or happenings has a huge impact on how safety behavior can develop. Communication plays a central role in sensemaking and management plays a crucial role in this.

Case 1

In this case, it is crucial how the management of the commissioner and the contractor reacts to the people who created the incident. Do they see it as a shortcoming of themselves or as a shortcoming of the operator and the mechanics? Do they address issues of conduct with an employee of another company, knowing that this is more difficult than addressing the same issue with one's own colleagues? Do they present the results of the incident analysis as an explanation of what went wrong or as an opportunity to learn for the future?

9.6 MANAGING RULES AND REGULATIONS WITHIN AN ORGANIZATION

Level 2 safety management is based on creating rules and regulations to control behavior. This way of managing behavior has proven to be very successful. Good rules are like condensed knowledge: Everything we know about a certain task or tool is integrated in an instruction on how to do a certain task or job. If one understands the rule, one understands a lot about the ins and outs of that aspect of work.

There are, however, serious indications that the amount of rules and regulations sometimes has gone beyond a point where they still are contributing to safety. Some claim that there are areas in which rules already have a counterproductive effect. There are several problems involved.

- The first problem with rules is that people can only digest a certain amount of them. Organizations have ring binders full of rules and regulations and expect that everybody knows them by heart.
- Another problem is that the more rules that are created, the greater the chance that different rules will start conflicting with each other, especially when different sources are involved (for example, government and company policy).
- From a psychological perspective, a major problem is that rules create a fake sense of safety. People rely on rules and expect that everybody sticks to them regardless what happens. Rules can bring us into a comfort zone mode that leads to not really paying attention to what's happening around us. In other words, rules wrongfully create a sense of safety that kills our safety alertness. We lose the necessary vigilance to monitor all our surrounding traffic.

Research in the area of traffic safety has recently shown that each situation has an optimum level of rules and regulations. One would expect that the more behavior is regulated and people stick to the traffic rules, the fewer accidents will happen on the street. Recent experiments with the Shared Space concept, developed by Monderman, show the opposite. Monderman states that the fewer rules we have and the more we have to rely on each other, the more socially we behave. If a driver is not sure what other road users are going to do, she has to watch them. Unpredictability raises the general anxiety level and activates safety awareness. Road users stay more vigilant if they are not fully sure how the other users are going to behave. They must all keep scanning the direction and speed of other road users and adjust their own behavior to it. This alertness has a stronger effect on safety than the predictability created by regulations of behavior. As stated before, we are very good in following the safety behavior of others. As long as the environment alerts us to do so, we will focus on others and anticipate their behavior. The fact that road users take more care in the above situation is also a reflection of the average speed, which is automatically reduced without physical obstacles like speed ramps.

More vigilance and safety with fewer rules.

Those who visited London 10 years ago could see small fences along the streets everywhere; they prevented pedestrians from suddenly crossing the street. In streets with moderate traffic (less than 8000 cars a day) the number of accidents with injury was reduced to almost 50 percent after removing these small fences (Webster, 2007). Above the 8000 cars a day, it is safer to separate the lanes and to disentangle the routes of the different road users. Many local communities in Sweden, Denmark, The Netherlands, France, and Macedonia have applied this principle when organizing their public space (Hammerin, 2006; Palmblad, 2008).

The general conclusion is that fewer accidents happen in situations where different streams of traffic are mixed in such a way that everybody easily notices that the streams are mixed.

Although traffic is a specific area within safety management, this concept of Shared Space shows that wherever people need to align their activities to those of others, creating a need for understanding behavior of others is much more effective than regulating behavior via rules. The traffic discipline is too specific for drawing general conclusions for safety management in general, but it definitely shows that more rules are not always better. Each situation probably has its own optimal level of regulations.

9.7 CORPORATE SAFETY PROGRAMS BASED ON PRIMING

Apart from safety buddies, team interventions, and management actions, an organization can also influence safety behavior via general communication and specific assignments for all employees. For this purpose, we introduce the technique of priming. Although priming is still hardly applied in organizational development, it has established an important place among the techniques of recent psychological research. Priming works completely nonconsciously and can be defined as a form of influencing behavior by exposing the brain to stimuli that are strongly connected to a stereotype or a certain stimulus. Priming effects can be observed at almost all stages of cognitive processing (Bargh, 2001; 2006; Dijksterhuis, 2005, Naccache & Dehaene, 2001).

A first impression about priming is given by a research from Bargh in 1996 (Dijksterhuis, 2007). He invited people for a whole morning of psychological tests. Test 1 was introduced as a language test and consisted of filling in a crossword puzzle. One group received a puzzle with words that all related to elderly people (*retired, pension, Florida, relaxed, grandchildren*, etc.). The other group received the same puzzle but with words that related to young people (*diploma, ambition, career, New York, first salary*, etc.). After they had finished the puzzle they were asked to go to another room, five floors lower, for a second test. The time between leaving the first room and entering the second room was clocked via a laser beam on both doors. The members of the subgroup who had been confronted with words connected to elderly people took significantly more time to bridge the distance between the both rooms, compared to the subgroup that had been given puzzles with words about young people. When checked later, neither of the groups had noticed that all the words in the puzzle were related to a certain topic.

This experiment shows the essence of priming. When one is presented with a strong stereotype, this has a nonconscious influence on the way we perform our behavior. The modern brain has a free working memory that can be filled with information for a certain task. Information has to be gathered from storage rooms that are spread over the brain, and it uses search engines to reduce the number of possibilities. The better those engines work, the more effectively we can feed our reasoning process. Selective searching is organized by using filters, and stereotyping is one of the supporting tools in this process. What is interesting is that basic stereotypes have an effect that lasts longer than the task for which they are required or activated. In fact, subsequent tasks will also be influenced by strong stereotypes used in previous tasks. In addition, stereotypes not only play a role in gathering information, they also influence what we do with the information. Stereotypes have an impact on quality and quantity of behavior. It is possible to use the concept of safety or a safe way of working as a stereotype for priming.

Priming is a form of influencing behavior by exposing the brain to stimuli that are strongly connected to a stereotype.

In the last few years, many forms of priming have been developed. Priming can be done by adding objects but also by triggering people via small tasks or activities. A picture of Einstein in a waiting room results in higher scores on a subsequent IQ test, and money makes us more task driven but also less social. Priming works intuitively as long as people can identify with the primed model. There is also evidence that priming can have a counterintuitive effect as soon as there is a felt distinction with the model (Aarts & Dijksterhuis, 2002). The experiment with Einstein in the waiting room led to poorer performance on an IQ test if university professors did this test. The picture of Einstein made them feel stupid, so they scored lower. When people were primed with really fast animals, they all walked more slowly because they felt their own slowness while watching those fast animals.

Priming operates on the nonconscious level, but some attention is needed during the priming phase (Naccache, Blandin, & Dehaene, 2002). People usually don't know that they are manipulated, and they also don't realize that they start to act in a specific way related to the stereotype. But even if people do know, the effect remains the same. You can even manipulate yourself and it will still work.

The word *priming* stems from painting and refers to the creation of a first layer of paint that has an influence on the attachment, the color, and the structure of the next layers. Priming as a psychological principle expresses that the first behavior (of a day or a session), and the stereotype behind it, influence and direct future behavior in a certain way. Priming is used extensively in psychological research as a way of provoking a certain mental condition (for example, secure, proud, ambitious, confident, etc.).

There is substantial evidence that priming via strong stereotypes can influence executive functions (for example, planning, desiring) for at least four hours. Priming twice a day can enhance safer behavior. If real people are involved in inducing priming, these people actually have to believe in the primed message; otherwise, the priming has no effect (Doyen, 2012).

It is questionable whether priming can also attribute to long-lasting changes in behavioral patterns. The most relevant variable in achieving this long-term change is the frequency and persistence of the influencing impulses. It is like supporting a newly planted young tree with a small pole. The support needs to be there until the tree is strong enough to withstand the winter storms. Once the tree has strong roots, the pole is no longer needed. A permanent series of primers can work as a supporting pole, helping until such behavior is sufficiently established.

The possibilities created by priming are enormous. The idea is that the employee be exposed to safety stereotypes before he starts working, at least twice a day. These stereotypes can be any of the following:

- Posters with safety-related issues at the entrance of the building or spread over the site
- Safety items as first topic in all meetings (normal, Toolbox, or evaluations) and presentations
- The LMRA (Last Minute Risk Assessment)

- Safety as a preamble in all plans, Individual Performance Management, and so on
- Safety discussions with team members
- Safety games and competitions
- Safety campaigns in changing forms, new speakers, new subjects
- Safety equipment and tools (for example, sunblock)

Priming is more effective if the stereotypes are connected to our evolutionary tasks: survival, group behavior, and reproduction. Our deepest concerns are probably those for our family, especially our children. Motivation for safety behavior can easily be generated here. The chemical company Huntsman once printed a picture of the employee's own family on the reverse of his personal company badge with the text: "This is the reason why we work safely." If really tough messages have to be communicated (in case of persistent casualties), a picture of a waiting mother and children can wake up employees. Humor can also be a good way to catch attention as long as safety itself is not made ridiculous.

Safety priming can and should be done via all senses. The most recent discovery is the use of odors. There is evidence that the smell of lemon and soap enhances a more clean way of eating. Further evidence shows that the smell of lemon has a positive effect on safe car driving, and the combined smell of citrus has an effect on safer behavior in general. The odor of peppermint led to a significant performance improvement in difficult motor tasks (Ho & Spence, 2005) and to more alertness in a visual attention task (Warm et al., 1991; Moss et al., 2008) it also helps to fight daytime sleepiness (Norrish, 2005). The impact of olfactory sense is special because it is the only sense that has a direct contact with the emotional brain, the source of our motivation. Due to this direct line, the translation of sensory input into action cannot be merged while processing concepts, words, or opinions. This makes smell a strong source of manipulation.

The smell of citrus is a known primer for safe driving. It works even if the scent is under the threshold of conscious perception.

9.7.1 Overdoses of Priming?

One might wonder whether there is an optimal amount of priming. Considering the effects of habituation, priming might go so far that it starts to lose its effectiveness. Current research indicates that regular priming tends to be effective as long as the actual stimuli change a little bit, although the message can be the same. Regular renewal of posters in a safety campaign is a good way to maintain the effect. In fact, the more familiar the campaign becomes, the more the employees start to like it. This principle is called the mere exposure effect (Zajonc, 2001). This effect works even if the stimuli are not perceived on a conscious level (for example, posters when entering the gate that are not explicitly watched because employees are passing by fast). The effect is bigger however if there is a conscious recognition (Newell, 2006).

9.7.2 Can Priming be Counterproductive?

Unfortunately, the principle of priming also works on unplanned occasions. Without realizing it, this principle is also active in situations in which nobody has any intention to influence or manipulate behavior in a certain direction. If a few employees meet at the coffee machine or water cooler and complain about work, other teams, or the employer, these complaints can influence the way of working for hours.

Case 1

In this example, we can discover one moment of priming, answering the questions on the LMRA and filling in the card.

9.8 SUMMARY

Priming is a mysterious but effective way to nonconsciously influence behavior of others. It is widely used in psychological research but hardly practiced in labor psychology. The concept is simple: a stereotype input (for example, words, images, smells) influences behavior for hours in a way according that stereotype. Smell has a direct entrance in the emotional system and evokes behavior more connected to our first nature. The direct manager is perceived as a role model. By doing what he expresses and expressing what he is doing, he can use the power connected to this perception. From a safety perspective, stress management means creating an optimum level of stress and avoiding both low and high stress levels. The willingness to take risks and trespass rules is strongly related to the expected outcome of a project. Especially in cases where the deadline might be missed, high risk behavior might be evoked. Investigating safety incidents with an understanding attitude, creates an atmosphere in which openness and improvement foster. There are an optimal amount of rules that vary per situation. More rules are seldom better.

TIPS FOR TRANSFER

Tip 1: Working Alone Versus in a Team

The saying "shared fear is half fear" expresses the principle that sensitivity goes down while having company. Also anxiety levels go down while operating together. There are many situations in which this companionship of another person is needed at work, for example, when extra hands are required, when complementary competences are involved or when one has to do the operation and the other has to check it.

Tip: In all cases in which there is doubt whether two or more persons are needed, consider giving the task only to one person. This person probably will stay more alert during the execution of the task.

Tip 2: Make Safety Sexy

The brain has a love–hate relationship with safety. Safety measures have to ensure that we stay sound and alive but shouldn't cost too much.

We can influence this image by making safety sexier by using fashionable safety equipment for example, all personal protection being Nike safety shoes, Ray Ban safety glasses, Armani overalls, and so on. As soon as employees are tempted to wear them for private usage, you are on the right track.

Question: What can you do to give safety an attractive appeal?

Tip 3: Reducing the Self-Attribution Failure

Perception is personal; everybody perceives something else. Although we have no options to compare what persons actually perceive, we have strong indications that some perceptions are more similar among persons than others. Our self-image belongs to the area in perception where we assume the most personal coloring of the image. As the self-image is important in matching competencies to activities and in managing safety margins, a correct estimation of own competencies and safety behavior is important.

Regular feedback about actual safety behavior can help to adjust the self-image so that it becomes more in line with what others perceive.

Both feedback from peers and from a safety buddy can help in this process.

Tips:

- Organize personal feedback moments on exposed safety behavior, for example, when a project is being evaluated.
- Manage this feedback at least a few times a year
- Incorporate safety behavior as an integral item in job appraisals.

Tip 4: Workshop Stress Management for Managers: How to Regulate Stress.

A usual course in stress management aims at reducing stress. Brain Based Safety states low stress is as dangerous as high stress. During high stress, our motor skills weaken and our perception narrows, but during low stress our system is just not ready to act, and our perception misses relevant details because it is lazy.

Stress management means managing stress to an optimum level: enough work to feel a light pressure, not so much work that one has problems finishing it within the demanded context.

Tip: Investigate the present stress level at work by observing, and make a plan to tackle both ends of the stress spectrum.

Tip 5: How to Deal with Safety Regulations.

There is no professional area in which there are no rules and regulations. Either government regulations dictate rules and regulations so that a company receives a license to operate, or company policy imposes them in order to smoothen and connect work processes. In some cases, the number of rules is enormous and rules might even be contradictory. This can create a tendency to neglect some of the rules. Often rules are seen as obligations that slow down work, sometimes without personnel even knowing reasons for the rules. The less clear a rule is, the higher the tendency to break it.

Every rule is made to reduce risks, and behind each rule there is often a world of hidden information. Rules usually are the result of thorough thinking, but the reasoning behind the rule is seldom available anymore.

Explaining rules to newcomers not only helps to improve the quality of the acceptance but also enhances the understanding of processes. Eventually, this also leads to improved risk understanding.

The advice to a safety coach who teaches the rules to a newcomer is to not only explain the rules but also discuss the reasons behind them. Is the safety coach practicing this insight in the way induction is organized in his or her organization?

Tip 6: Project Priming Safety

We know that the effects of priming can last at least four hours and that continuous priming can lead to behavioral changes. If you decide to use the principles of priming to enhance safety in your organization, it is wise to develop an overall master plan. Possible elements of such master plan are listed here:

- A new name for the safety campaign
- A logo that can be used (connection to company and safety)
- A safety slogan
- A plan that covers more than one year, in which it is possible to change the main topic from time to time (Safety first, Safety together, Safety at home, Personal protection, Challenge me; challenge you)
- Each topic is launched with a kick-off session
- A multimedia approach with posters, billboards, flat screen instructions, pop-ups on PCs, mails, articles in the company magazine, small films posted on You Tube and distributed via intranet
- A project plan in which there is attention for safety per half-day, per week, per month, per quarter, and per year:
  - An exposure to a safety message at least every four hours with care taken that this exposure changes from time to time (place, timing, form)
  - Attention for safety in every magazine, intranet news, and so on
  - Safety as first item in every meeting
  - Quarterly meetings organized around complete safety
  - A billboard with safety figures, like the Safety Index (see also Chapter 10), OSHA rates, and so on
- Creation of a new safety character (for example, Safe Simon) and a counterpart (Dangerous Danny) who visualize how to act safe (or not) at critical moments

- Ways to challenge employees daily for three minutes, preferably at the start of the shift, for example, a Safety Quiz where you (and your team) can earn Safety Miles (offer the prize winners a trip)
- Organization of small sessions (funny and serious) with Safe Simon on unexpected moments in the plant
- Starting every meeting with safety issues related to the assignment
- Experimenting with fragrance, especially citrus or lemon

Tip: Write down some first ideas here, and make a project plan later.

Remember: Priming is relatively cheap and proven effective!

PART 4

Organizational Safety Management

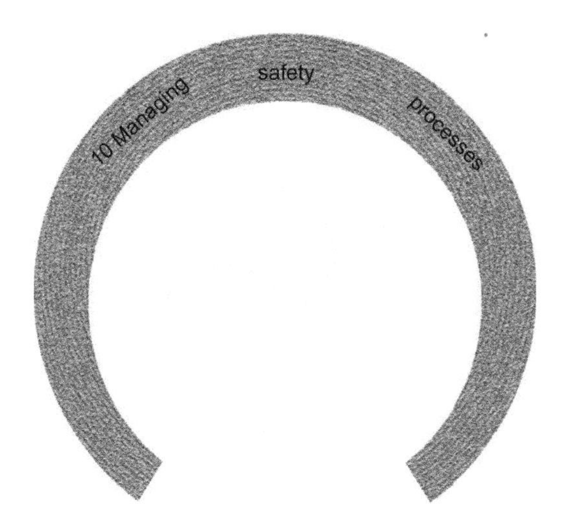

This final chapter is about managing processes directly and indirectly related to safety performance. First, we discuss the monitoring system of safety management and then how safety behavior can be rewarded.

CHAPTER 10

How to Manage Safety in an Organization

10.1 MONITORING SAFETY

In general, a closed feedback loop is necessary to monitor processes. Monitoring is a key precondition of managing. One of the major problems presently with safety management is that we mainly monitor in an old-fashioned way. Recorded incidents are mostly combined with subgroups of injuries, their treatment, and their results in lost time. If we examine the basics of management by objectives, we remember the mnemonic SMART, (Specific, Measurable, Acceptable, Realistic & Traceable). Traceable means that we can manage only if the path has enough breadcrumbs to mark it. Without these breadcrumbs, we have too little information to manage and we lose track. In safety management, the recorded incidents are usually the breadcrumbs. The more successful safety management becomes, the fewer incidents, the less breadcrumps, and the more difficult it becomes for the track to be followed and monitored.

We can only manage processes that deliver regular feedback.

If a plant has only a few recordable incidents per year, a new way of monitoring is required. This way must generate more information more regularly that can be used as feedback. The idea is to develop a Safety Index, an indicator of the quality and quantity of the actual level of safety management within an organization or specific department. The Safety Index consists of indicators of all the elements of a process: input, throughput and output. Some suggestions on new ways of monitoring follow:

- Considering the output, apart from the normal registration of the incidents, the amount of absence due to work-related illness (including stress, conflicts, etc.) can be a first additional indicator.
- The next step to enlarge the output indicator is to shift attention away from recorded incidents to near hits and to employee hazard notifications (Herbertson, 2008). Going one step further, a register of irregularities can be compiled, that is, things that are not the way they should be but have not yet directly led to hazards. If all data are gathered, it is possible to relate the number of near hits to the number of hazardous notifications and the number of irregularities. These data together resemble the safety pyramid of Heinrich (1931, quoted in Hollnagel 2009), but now on the more invisible part of safety management.
- A desirable side effect of recording near hits is that the focus of middle management will also be redirected to this subject.
- Throughput indicators can be generated via an employee survey which creates more indirect figures like safety perception of management or employees. Send out a short survey every month or every quarter to one-twelfth or one-quarter, respectively, of all the employees, in such a way that each employee answers a questionnaire once a year. This achieves good results, and in this way the organization receives regular feedback.
- Another throughput indicator is the results of audits. Many companies have a regular SHEM audit these days, in which all safety-related elements like critical procedures and safety updates of installations are surveyed. Other throughput indicators are the amount of Toolbox meetings and LMRA's.
- On the input side we can collect indirect variables like the number of hours allocated to safety training, the number of people involved, the percentage of employees that have a diploma in basic safety, the number of safety-related articles in the company magazine, the number of safety-related items on the agenda's of management meetings, and the number of priming activities as discussed in Chapter 9, etc.

The consolidation of all these figures together can generate a value, a Safety Index.

10.2 REGRESSION EFFECTS

For those who plan special safety programs in specific parts of the organization it is wise to take into account that safety processes seldom are stable and show fluctuations. The Safety Index will go up and down, usually without our understanding why.

Suppose there are 10 departments and that three of them score very badly on the Safety Index. The first idea would be to start a special program to upgrade the safety situation of those three departments. Suppose that a year later, after a special program, the Safety Index of these three departments has improved. They are no longer the safety laggards of the company but have an average score.

Initially, we might believe that the special program has been successful. Each of the three departments has improved, one more than the others. But in theory the improvement could be fully related to normal fluctuations. Statistically, we talk about regression to the mean, a way of saying that a lot of processes (like the weather) fluctuate during the years around a certain mean, and a good year will probably be followed by a bad year, and vice versa. Regression to the mean implicates that the Safety Index of those three departments that scored low would rise anyway just because of these fluctuations. A higher score then would have no direct meaning. Only if the Safety Index stayed high for many years would the option that the Index has structurally improved gain credence.

The basic message is this: Beware of evaluating special safety projects too fast, in order to avoid unrealistic conclusions about the effectiveness of such programs.

10.3 HR AND SAFETY: REWARDING SAFETY BEHAVIOR?

A final word concerning the reward system: It is not uncommon that organizations work with financial bonuses and that the range of the bonus is to some extent related to safety performance. Bonuses are related to extrinsic motivation that can be used for all behaviors people don't want to engage in for their own interest but will cooperate with because someone else appreciates these behaviors. Bonuses are effective at the instigation of safety management, especially on the level of rules, regulations, and protocols, because bonuses can help employees stick to these regulations. The Brain Based Safety approach for enhancing organizational safety has a basic assumption: Safety can only be realized when it is endemic to our total system and integrated in every brain function. It belongs to the domain of intrinsic motivation. What will happen if an organization starts to motivate people or managers on issues that belong to the intrinsic domain via extrinsic motivators like financial bonuses to attain higher levels of safety behavior? Will intrinsic and extrinsic motivation "match up" with each other? Will they strengthen or destroy each other?

Deci, Koestner, & Ryan (1999) state, "Tangible rewards ... tend to forestall self regulation. In other words, reward contingencies undermine people's taking responsibility for motivating or regulating themselves." Mellström and Johannesson (2008) showed that a financial reward for becoming a blood donor decreased significantly the readiness to become such a donor. Both theories rely on a phenomenon called "crowding out". Essential to this theory is the attribution of meaning to behavior. As soon as one receives an external incentive to act safely, safety enters a different normative domain. Safety no longer finds its origin in a deeply felt responsibility for one's own and others' well-being. It becomes an appreciated behavior carried out to please somebody. Crowding out states that extrinsic motivators transform behavior that is founded in a sense of responsibility or community into a behavior that is carried out for personal interest. A comparable study on priming with money shows similar results. If people are primed with concepts of money (for example, an expensive bag, a silver pen, or a screensaver with images of money), they slip into a self-supporting mode and forget others (Vohs, Mead, & Goode, 2008). They stop caring about the well-being of others and also forget to seek help for their own problems. Suddenly, tasks become a solitary pursuit. To summarize, a bonus system will work if an organization is still improving safety at a level of rules and regulations; it will have an opposite effect if the focus is on enhancing deeply held beliefs of safety behavior.

> Processes driven by extrinsic motivation can be reinforced by extrinsic rewards like bonuses. Processes driven by intrinsic motivation can only be reinforced by creating sense.

Previously, we have seen that safety is heavily reliant on social contact within an organization. Employees support and correct each other in order to create a safe environment. As soon as the introduction of money changes this process, a fundamental pillar of Brain Based Safety collapses. The results of safety management cannot be extrinsically motivated and will decrease if done so under such circumstances.

One might ask, what will happen if a bonus system based on safety targets is cancelled due to this new insight? Performance will probably decline initially, because the extrinsically motivated employees who acted safely partly because of the financial rewards involved will direct less attention to safety if the bonus shifts to other areas of work. But soon colleagues in the work environment will become disturbed because someone is not strictly following the rules and paying attention pro-actively to safety. Not paying attention to safety is not only gambling with your own life but also with the lives of others. A second reason why the safety performance might go down is that as soon as the bonuses are cancelled, there is no longer any reason to conceal near misses or minor incidents from management. So the Safety Index will probably become a more realistic indicator of the actual safety situation. After this process, performance will gradually increase as safety becomes intrinsically motivated.

10.4 SUMMARY

In order to monitor safety, we need more figures than just the registered incidents and near misses. Employees' perception of safety is valuable because employees are strong influencers of the actual safety performance. Audits will complete the total picture with an outsider's perspective. Financial bonuses stimulate people but seduce them to act in a more solitary way and for their own benefit. As safety is created with and for each other, intrinsic motivation is a much more suitable reinforcer than money.

TIPS FOR TRANSFER

Tip 1: Develop a Safety Dashboard Leading to a Safety Index

Can you design a dashboard or a balanced scorecard that gives an actual indication of the safety performance/situation at a certain moment? A dashboard works well if it allows the management to take action if the figures are not satisfying.

Question: What should be registered?

Suggestion: Can you integrate indicators of risk sensitivity, risk awareness, safety alertness, and safety decisions into the dashboard?

Question: How can you measure safety indicators in a way that is realistic, acceptable, and affordable?

Suggestion: Combine the measures via a weighted score into a Safety Index.

Tip 2: Safety Inspections

Many organizations have something like a daily inspection round, and in many cases safety is one of the items that is inspected. As soon as employees have finished their induction phase, they can be qualified for doing safety inspections.

Tips:

- Involve as many employees as possible to do safety inspections. It helps to call attention to safety in a positive way.
- Compare the safety reports of different safety inspectors so that they learn from each other how to investigate safety.
- Use the night shift for technical safety inspections. It keeps the inspectors awake and alert and contributes to safety awareness.

Tip 3: Bonus: Not a Financial Reinforcer for Safety

Bonuses are introduced for stimulating employees to do the right things and to spend extra effort on these right things. Money enhances extrinsic motivation. Unfortunately, it also kills intrinsic motivation and cooperation. It is wise to reconsider a bonus system for safety behavior. Within the theory of Brain Based Safety, a bonus for safety is like cursing in church. It shows no respect for physical integrity and it denies fundamental laws of behavior. If one really wants a bonus system, it is wise to organize it for targets in which no mutual cooperation is needed and which demand external motivation.

Never position safety management and safety behavior as an item that is done to please an employer, but frame it as care for employees.

By the way, a positive effect of the introduction of an individual bonus system on the performance of an organization never has been proven. The positive effect of incidental individual progress is compensated by the negative effect of a loss of cooperation. Bonuses only have an effect on the production within simple tasks that involve a lot of repetition.

SAFETY PHILOSOPHY

Safety in and of itself does not exist; people can only *act* safely.

It is possible to organize work in such a way that everybody can finish work in the same physical shape as he or she started it.

It is possible to organize private life in such a way that everybody can arrive at work in the same physical shape as he or she left it after the last shift.

It is to be expected that others commit themselves to make my work and life safer.

It is to be expected that I commit myself to make the work and life of others safer.

There is no valid reason why interventions to enhance safety should not be carried out.

There is no place where safety is not required.

Safety is the most important objective to attain. If we can't do it safe, it is better not to do it.

Safety is not a value—it is a virtue; we have to live it.

This page intentionally left blank

BIBLIOGRAPHY

Aarts, H., & Dijksterhuis, A. (2002). Category activation effects in judgment and behaviour: The moderating role of perceived comparability. *British Journal of Social Psychology* 41:123–138.

Alexander, W. H., & Brown, J. W. (2011). Medial prefrontal cortex as an action-outcome predictor. *Nature Neuroscience* 14:1338–1344.

Bargh, J. A., Gollwitzer, P. M., Lee-Chai, A., Barndollar, et al. (2001). The automated will: non-conscious activation and pursuit of behavioral goals. *Journal of Personal and Social Psychology, Dec* 81(6):1014–1027.

Bargh, J. (2006). What have we been priming all these years? On the development, mechanisms, and ecology of nonconscious social behavior. *European Journal of Social Psychology* 36(2):147–168.

Barbeau, E., Wendling, F., Régis, J., Duncan, R., Poncet, M., et al. (2005). Recollection of vivid memories after perirhinal region stimulations: Synchronization in the theta range of spatially distributed brain areas. *Neuropsychologia* 43:1329–1337.

Berwick, D. M. (2003). Disseminating innovations in health care. *Journal of the American Medical Association* 289(15):1969–1975.

Carey, B. (2009). In battle, hunches prove to be valuable. *New York Times* July 27.

Chaiken, S., & Trope, Y. (1999). *Dual-process models in social psychology*. New York: Guilford.

Chandler, J. J., & Pronin, E. (2011). Fast thought speed induces risk taking. *Psychological Science* 23(4):370–374.

Chaumon, M., Hasboun, M., Baulac, M., Adam, C., & Tallon-Baudry, C. (2009). Unconscious contextual memory affects early responses in the anterior temporal lobe. *Brain Research* 1285:77–87.

Daalmans, J. M. T. (2011). *De breingids*. Amsterdam: Boom Uitgeverij (the brain guide).

Damasio, A. R. (1995). *Descartes' error—Emotion, reason and the human brain*. New York: Putnam's Sons.

Deci, E., Koestner, R., & Ryan, R. (1999). A meta-analytic review of experiments examining the effects of extrinsic rewards on intrinsic motivation. *Psychological Bulletin* 125:627–688.

DeJoy, D. M. (1994). Managing safety in the workplace: An attribution theory analysis and model. *Journal of Safety Research* 25(1):3–17.

Devlin, J. T., & Price, C. J. (2007). Perirhinal contributions to human visual perception. *Current* 17(17):1484–1488.

Dijksterhuis, A. (2007). Het slimme onbewuste. Amsterdam: Bert Bakker (The clever unconsciousness).

Dijksterhuis, A., Aarts, H., & Smith, P. K. (2005). The power of the subliminal: On subliminal persuasion and other potential applications. In R. R. Hassin, J. S. Uleman, & J. A. Bargh (eds.), *The new unconscious* (pp. 77–106). New York: Oxford University Press.

Dobelli, R. (2011). Die kunst des klaren denkens. München: Carl Hanser Verlag (The art of clear thinking).

Duncan, S., & Barrett, L. F. (2007). Affect as a form of cognition: A neurobiological analysis. *Cognition and Emotion* 21(6):1184–1211.

Epstein, S. (1994). Integration of the cognitive and the psychodynamic unconscious. *American Psychology* 49:709–724.

Galbraith, G. C., Arbagey, P. W., Branski, R., Comerci, N., & Rector, P. (1995). Intelligible speech encoded in the human brain stem frequency-following response. *Neuroreport* 6(17):2363–2367.

Gilbert, D. T., & Malone, P. S. (1995). The Correspondence Bias. *American Psychological Association* 117(1):21–38.

Granovetter, M. (1973). Strength of weak ties. *American Journal of Sociology* 78:1360–1380.

Hammerin. (2006). From Part 3, p. 38.

Hatzigeorgiadis, A., Zourbanos, N., Galanis, E., & Theodorakis, Y. (2011). Self-talk and sports performance, a meta-analysis. *Perspectives on psychological science* 6(4):348–356.

Herbertson, W. G. (2008). *The practical safety guide to zero harm. How to effectively manage safety in the workplace.* Australia: The Value Organisation Party Ltd.

Heuvel, P. v. d., & Sporns, O. (2011). Rich-club organization of the human connectome. *The Journal of Neuroscience* 31(44):15775–15786.

Ho, C., & Spence, C. (2005). Olfactory facilitation of dual-task performance. *Elsevier Neuroscience Letters* 389:35–40.

Hollnagel, E. (2009). *Safer complex industrial environments: a human factors approach.* CRC Press.

Hyman, I. E., Boss, S. M., Wise, B. M., McKenzie, K. E., & Caggiano, J. M. (2009). Did you see the unicycling clown? Inattentional blindness while walking and talking on a cell phone. *Applied Cognitive Psychology* 24697–607

Iacoboni, M. (2008). *Mirroring people—The new science of how we connect with others.* New York: Farrar, Straus and Giroux.

Kahneman, D. (2011). *Thinking, fast and slow.* New York: Penguin.

Keyser, C., & Gazzola, V. (2006). Towards a unifying neural theory of social cognition. In S. Anders, G. Ende, M. Junghöfer, J. Kissler, & D. Wildgrüber (eds.), *Progress in brain research: Understanding emotions* (vol 156, pp. 379–401). Amsterdam: Elsevier.

Krishnan, A., Xu, Y., Gandour, J. T., & Cariani, P. A. (2004). Human frequency-following response: representation of pitch contours in Chinese tones. *Hearing Research* 189:1–12.

Kokal, I., Gazzola, V., & Keysers, C. (2009). Acting together in and beyond the mirror neuron system. *Neuroimage* 47(4):2046–2056.

Kolb, B., & Whishaw, I. Q. (2008). *Fundamentals of human neuropsychology.* New York: Freeman and Company.

Kucharczyk, E. R., Morgan, K., & Hall, A. P. (2012). The occupational impact of sleep quality and insomnia symptoms. *Sleep Medical Review* Epub ahead of print.

Kühn, S., Müller, B. C. N., van Baaren, R. B., Wietzker, A., Dijksterhuis, A., et al. (2010). Why do I like you when you behave like me? Neural mechanisms mediating positive consequences of observing someone being imitated. *Social Neuroscience* 5(4):384–392.

Ledoux, J. (2002). The emotional brain, fear, and the amygdala. *Cellular and Molecular Neurobiology* 23(4/5):727–738.

Lamme, V. (2010). *De vrije wil bestaat niet.* Amsterdam: Bert Bakker (there is no such thing as a free will).

Mack, A. (2003). Inattentional blindness: Looking without seeing. *Current Directions in Psychological Science* 12(5):179–184.

MacLean, P. D. (1990). *The triune brain in evolution: role in paleocerebral functions.* New York: Plenum Press.

Mellström, C., & Johannesson, M. (2008). Crowding out in blood donation: Was Titmuss right? *Journal of the European Economic Association* 6(4):845–863.

Mobbs, D., Marchant, J. L., Hassabis, D., Seymour, B., Tan, G., et al. (2009). From threat to fear: The neural organization of defensive fear systems in humans. *The Journal of Neuroscience* 29(39):12236–12243.

Moss, M., Hewitt, S., Moss, L., & Wesnes, K. (2008). Modulation of cognitive performance and mood by aromas of peppermint and ylang-ylang. *Journal of Neuroscience* 118(1):59–77.

Most, S. (2010). What is "inattentional" about inattentional blindness. *Consciousness and Cognition* 19(4):1102–1104.

Murray, E. A. (2007). The amygdala, reward and emotion. *Trends in Cognitive Sciences* 11(11):489–497.

Naccache, L., & Dehaene, S. (2001). The priming method: Imaging unconscious repetition reveals an abstract representation of number in the parietal lobes. *Cerebral Cortex* 11:966–974.

Naccache, L., Blandin, E., & Dehaene, S. (2002). Unconscious masked priming depends on temporal attention. *Psychological Science* 13(5):416–424.

Newell. (2006). From Part 3, p. 41.

Norrish, M. I., & Dwyer, K. L. (2005). Preliminary investigation of the effect of peppermint oil on an objective measure of daytime sleepiness. *International Journal of Psychophysiology* 55:291–298.

NSWO (Nederlands Slaap/Waak Onderzoek). (2012). *Werk houdt veel Nederlanders uit hun slaap.* NSWO.nl (work keeps many Dutch people awake).

Pessoa, L. (2008). On the relationship between emotion and cognition. *Nature Reviews Neuroscience* 9:148–158.

Pessoa, L., & Adolphs, R. (2010). Emotion processing and the amygdala: From a "low road" to "many roads" of evaluating biological significance. *Nature Reviews Neuroscience* 11(11):773–778.

Petrovic, P., Kalso, E., Petersson, K. M., & Ingvar, M. (2002). Placebo and opioid analgesia—imaging a shared neuronal network. *Science* 295:1737–1740.

Rizzolatti, G., & Craighero, L. (2004). The mirror-neuron system. *Annual Review of Neuroscience* 27:169–192.

Rogers, E. M. (1995). *Diffusion of innovations.* 4th ed.New York: Free Press.

Root, J. C., Tuescher, O., Cunningham-Bussel, A., Pan, H., Epstein, J., et al. (2009). Frontolimbic function and cortisol reactivity in response to emotional stimuli. *NeuroReport* 20:429–434.

Sandi, C., & Pinelo-Nava, T. M. (2007). Stress and memory: Behavioral effects and neurobiological mechanisms. *Neural Plasticity* Article ID 78970.

Schippers, M. B., Gazzola, V., Goebel, R., & Keysers, C. (2009). Playing charades in the fMRI: Are mirror and/or mentalizing areas involved in gestural communication? *Plos ONE* 4(8):e6801.

Schippers, M. B., Roebroeck, A., Renkena, R., Nanettia, L., & Keysers, C. (2010). Mapping the information flow from one brain to another during gestural communication. *PNAS* May 3. Published online before print.

Sharot, T., Korn, C. W., & Dolan, R. J. (2011). How unrealistic optimism is maintained in the face of reality. *Nature Neuroscience* 14:1475–1479.

Shaver, K. G. (1970). Defensive attribution: effects of severity and relevance on the responsibility assigned for an accident. *Journal of Personality and Social Psychology* 14(2):101–113.

Sloman, S. A. (1996). The empirical case for two systems of reasoning. *Psychological Bulletin* 119:3–22.

Taylor, J. R., & Van Every, E. J. (2000). *The emergent organization: Communication as its site and surface*. Mahwah, NJ: Erlbaum.

Tiemeijer, W. L. (2011). *Hoe mensen keuzes maken. De psychologie van het beslissen*. Amsterdam: Amsterdam University Press (how people make choices. The psychology of deciding).

Thaler, R. (1980). Toward a positive theory of consumer choice. *Journal of Economic Behavior & Organization* 1:39–60.

Van kleef, G. A., Homan, A. C., Finkenauer, C., Gündemir, S., & Stamkou, E. (2011). Breaking the rules to rise to power: how norm violators gain power in the eyes of others. *Social Psychological and Personality Science* 2:500.

Vohs, K. D., Mead, N. L., & Goode, M. R. (2008). Merely activating the concept of money changes personal and interpersonal behaviour. *Current Directions in Psychological Science* 17(3):208–212.

Warm, J. S., Dember, W. N., & Parasurama, R. (1991). Effects of olfactory stimulation on performance and stress in a visual sustained attention task. *Journal of the Society of Cosmetic Chemists* 42(1991 May/June):199–210.

Weick, K. E., & Sutcliffe, K. M. (2005). *Managing the unexpected: Resilience performance in an age of uncertainty*. New York: John Wiley & Sons.

Weick, K. E., Sutcliffe, K. M., & Obstfeld, D. (1999). Organizing for high reliability: Processes of collective mindfulness. In R. S. Sutton, & B. M. Staw (eds.), *Research in organizational behavior* (vol. 1, pp. 81–123). Stanford: Jai Press.

Weick, K. E., Sutcliffe, K. M., & Obstfeld, D. (2005). Organizing and the process of sensemaking. *Organizational Science* 16(4):409–421.

Zajonc. (2001). From Part 3, p. 41.

INDEX

T

Printed and bound by CPI Group (UK) Ltd, Croydon, CR0 4YY
08/06/2025
01896868-0001